Praise for **Literary The...**

"Surprising, funny and resolutely unintim... has figured out how to present a web of con...
—Jennifer Szalai, *New York Times Book Review*

"Through his use of history and philosophy, Yi Tenen reminds us that technology cannot exist independently of its creators and users, and ultimately it is we humans who hold the final responsibility for technology's impact on the world. This is a book that makes you a little wiser in realizing that responsibility."
—John Warner, *Chicago Tribune*

"A deft, refreshing book about the truth of how technology develops: as the product of collective human history."
—Jesse Damiani, *Forbes*

"A witty, if challenging, read."
—Andrew Robinson, *Nature*

"A delightfully fresh perspective on AI. [Yi Tenen] looks back more than a thousand years into literary history to report on an astonishing variety of algorithmic strategies for writing poems, plays, novels, fables, and much more. . . . By combining his literary knowledge with a background as a Microsoft software engineer, Yi Tenen brings readers a human-centered way of appreciating technology. . . . [T]he final chapter provides valuable guidance for how to think about AI. . . . [A] deeply human endeavor that emerges from a long history and broad collaboration."
—Ben Shneiderman, University of Maryland

"Reading this book is like taking a walk in a literary forest. You will see up close trees whose names you never knew and discover paths that lead your mind in new directions. Yi Tenen guides us along the way, by putting in historical context how machines started out as voracious readers and emerged as creative writers."

—Jeannette M. Wing,
executive vice president for research and
professor of computer science, Columbia University

"*Literary Theory for Robots* is many things—brainy, chatty, charming, disarming—but, above all, it is great fun to read. Dennis Yi Tenen's cast of 'lovely weirdos' and their wheels, charts, templates, schemas, and links will stay with me for a long time. So will his insistence that intelligence is a social and collective phenomenon, one whose history reveals the human presence behind every machine." —Merve Emre, author of

The Personality Brokers: The Strange History of
Myers-Briggs and the Birth of Personality Testing

"*Literary Theory for Robots* serves as an alternative to the breathless utopian or apocalyptic hallucinations of the tech bros funding the AI revolution, instead offering a highly relatable perspective on thinking machines grounded in history, literature, and lived human experience. Yi Tenen shows that truly understanding the future of our digital augmentation depends not on more STEM but on more liberal arts. This book will be remembered as the moment thinking people realized how to raise better robots: read them good stories."

—Douglas Rushkoff, author of *Survival of the Richest:*
Escape Fantasies of the Tech Billionaires

"Intriguing. . . . Yi Tenen, stirring some wit and anecdotes into the story, sets out the material in non-technical terms, making for an entertaining, informative read. An eclectic and erudite tale of how wide-eyed visions become smart, interactive tools." —*Kirkus Reviews*

"Timely and original, this is an essential resource on the history of text generating AI, and its future." —*Publishers Weekly*

Literary Theory
for Robots

NORTON
SHORTS

Literary Theory
for Robots

How Computers Learned to Write

DENNIS YI TENEN

W. W. NORTON & COMPANY
Independent Publishers Since 1923

For information about permission to reproduce selections from this book,
write to Permissions, W. W. Norton & Company, Inc., 500 Fifth Avenue,
New York, NY 10110

For information about special discounts for bulk purchases, please
contact W. W. Norton Special Sales at specialsales@wwnorton.com or
800-233-4830

Manufacturing by Lakeside Book Company
Production manager: Delaney Adams

Library of Congress Control Number: 2024944779

ISBN: 978-1-324-10505-3 pbk.

W. W. Norton & Company, Inc.
500 Fifth Avenue, New York, N.Y. 10110
www.wwnorton.com

W. W. Norton & Company Ltd.
15 Carlisle Street, London W1D 3BS

10 9 8 7 6 5 4 3 2 1

CONTENTS

Literary Theory
for Robots

CHAPTER 1

Intelligence as Metaphor (An Introduction)

■　　■　　■　　■　　■　　■　　■　　■

COMPUTERS LOVE TO READ. And it isn't just fiction before going to bed. They read greedily: all literature, all of the time—novels, encyclopedias, academic articles, private messages, advertisements, love letters, news stories, hate speech, and crime reports—everything written and transmitted, no matter how insignificant.

This ingested printed matter contains the messiness of human wisdom and emotion—the information and disinformation, fact and metaphor. While we were building railroads, fighting wars, and buying shoes online, the machine child went to school.

Literary computers scribble everywhere now in the background, powering search engines, recommendations systems, and customer service chatbots. They flag offensive content on social networks and delete spam from our inboxes. At the hospital, they help convert patient-doctor conversations into insurance billing codes. Sometimes, they alert law enforcement to potential terrorist plots and predict (poorly) the threat of violence on social media. Legal professionals use them to hide or discover evidence of corporate fraud. Students are writing their next school paper with the aid of a smart word processor, capable not just of completing sentences, but generating entire essays on any topic.

In the industrial age, automation came for the shoemaker and the factory-line worker. Today, it has come for the writer, the professor, the physician, and the attorney. All human activity now passes through a computational pipeline—even the sanitation worker transforms effluence into data. Like it or not, we have all become subjects to automation. To survive intact, we must also learn to become part software engineers and part, well—whatever you are doing is great!

If any of the above comes as a surprise to you, my job, I feel, is mostly done. Curiosity piqued, you will now start noticing literary robots everywhere and join me in pondering their origins. Those not surprised perhaps believe (erroneously!) that these siliconites have learned to chat only recently, somewhere on the fields of computer science or software engineering. I am here to tell you that machines have been getting smarter in this way for centuries, long before computers, advancing on far more arcane grounds, like rhetoric, linguistics, hermeneutics, literary theory, semiotics, and philology.

So that we may hear them speak—to read and understand a vast library of machine texts—I want to introduce several essential ideas underpinning the ordinary magic of literary computers. Hidden deep within the circuitry of everyday devices—yes, even "smart" light bulbs and refrigerators—we will find tiny poems that have yet to name their genre. By the end of the book, I hope you can see this technology anew, replete not just with instrumental capacity (to keep food cold or to give light) but with potential for creativity and collaboration.

Throughout, we will be tempted to ask existential questions about the nature of artificially intelligent things. "How smart are they?" "Do they really 'think' or 'understand' our language?" "Will they ever—have they already—become sentient?"

Such questions are impossible to answer (in the way asked), because the very categories of consciousness derive from human experience. To understand alien life forms, we must think in alien ways. And rather than argue about definitions ("Are they smart or not?"), we can begin to describe the ways in which the meaning of intelligence continues to evolve.

Not long ago, one way of appearing smart involved memorizing a bunch of obscure facts—to become a walking encyclopedia. Today, that way of knowing seems like a waste of precious mental space. Vast online databases make effective search habits more important than rote memorization. Intelligence changes. The puzzle of its essence cannot therefore be assembled from sharp, binary attributes, laid out always and everywhere in the same way: "Can machines think: yes or no?" Rather, we can start putting the pieces together contextually, at specific times and places, and from the view of an evolving, shared capacity: "How do they think?" and "How do we think with them?" and "How does that change the meaning of thinking?"

In answering the "how" questions, we will discover a strange sort of a twinned history, spanning the arts and sciences. Humans have been thinking in this way—with and through machines—for centuries, just as they have been thinking with and through us. The mind, hand, and tool move at once, in unison. But the way we train minds, hands, or tools treats them almost like entirely separate appendages, located in different buildings, in unrelated fields on a university campus. Such an educational model isolates ends from means and means from ends, disempowering its publics. In this volume, I would like to imagine an alternative, more integrated curriculum, offered to poets and engineers alike—bound, eventually, for a machine reader as part of another training corpus.

Next time you pick up a "smart" device, like a book or a phone, pause mid-use to reflect on your body posture. You are watching a video or writing an email perhaps. The mind moves, requiring mental prowess like perception and interpretation. But the hand moves, too, animating body in concert with technology. Pay attention to the posture of the intellect—the way your head tilts, the movement of individual fingers, pressing buttons or pointing in particular ways. Feel the glass of the screen, the roughness of paper. Leaf and swipe. Such physical rituals—incantations manifesting thought, body, and tool—bring forth the artifice of intellect. The whole thing is "it." And that's already kind of the point: Thinking happens in the mind, by the hand, with a tool—and, by extension, with the help of others. Thought moves by mental powers, alongside the physical, the instrumental, and the social.

What separates natural from artificial forces in that chain? Does natural intelligence end where I think something to myself, silently, alone? How about using a notebook or calling a friend for advice? What about going to the library or consulting an encyclopedia? Or in conversation with a machine? None of the boundaries seem convincing. Intelligence demands artifice. *Webster's* dictionary defines *intelligence* as the "skilled use of reason." *Artifice* itself stems from the Latin *ars*, signifying skilled work, and *facere*, meaning "to make." In other words, artificial intelligence just means "reason + skill." There are no hard boundaries here—only synergy, between the human mind and its extensions.

What about smart objects? First thing in the morning, I stretch and, at the same time, reach for my phone: to check my schedule, read the news, and bask in the faint glow of kudos, hearts, and likes from various social apps. How did I get into this position? I ask alongside the beetle from Kafka's *The Metamorphosis*. Who taught me to move like this?

It wasn't planned, really. Nor are we actually beetles living in our natural habitat, an ancient forest floor. Our intimate rituals morph organically in response to a changing environment. We dwell inside crafted spaces, containing the designs for a purposeful living. The room says, *Eat here, sleep there*; the bed, *Lie on me this way*; the screen, *Hold me like this*. Smart objects further change in response to our inputs. To do that, they must be able to communicate: to contain a layer of written instructions. Somewhere in the linkage between the tap of my finger and the responding pixel on-screen, an algorithm has registered my preference for a morning routine. I am the input and output: the tools evolve as they transform me in return. And so, I go back to bed.

THE CHAPTERS OF THIS book correspond to what I see as the three major gaps in our collective thinking about artificial intelligence. First, I will argue that our understanding of AI can and must become more grounded in the history of the humanities. Second, that a software-engineering education necessarily involves a number of inescapable philosophical problems, related to several nerdy-but-I'm-sure-you'll-find-them-fascinating kinds of textual matters. Finally, I propose to consider AI as part of the broader social and political dynamics arising, more generally, from the automation of labor, particularly as it concerns intellectual labor or "knowledge work."

History will remain my first and primary concern. Insert something clever here about forgetting the past to the detriment of the future. Any field developing rapidly, as computer science has been, can be forgiven for orienting itself toward the future. But not for long. Yes, one can learn to type quickly or to write a spreadsheet macro without a second thought. Reflect, however, on the immense frustration of commandeering all of our clumsy, impoverished digital avatars. You want me to save a what, where? File? Folder? Server? Cloud? The pervasive presence of metaphors should hint at the extent of our collective alienation. What is a file? Wherefore a folder? Why are they even called that? Who thought of putting exactly eight bits into a byte, and why not nine? How does a keystroke transform itself into a configuration of pixels on screen, forming a letter? Explain a common algorithm used to search, finish sentences when typing, filter spam, or to recommend products. No way. Too much. Such questions fall out of serious consideration because, paradoxically, they are "too simple" for computer science and way, way too complex for baseline literacy.

But first, why do history? One can simply launch into the future from the state-of-the-art, the state-of-the-present. At least two points are needed to make a line, however. History's subtle advantage lies in the revealing of trajectories through time. AI and indeed "artifice" or "intellect" (and even the toasting of bread) represent a cluster of related objects, ideas, techniques, practices, and institutions. A static snapshot contains their anatomy in three dimensions—this is how modern toasters work and this is how a mind thinks today. History expands these arrangements into the fourth dimension of time. Like any technology, the mind changes over time. Its anatomical hardware stays the same, but the software receives regular updates: all those encyclopedias, calculators, search tools, and other smart devices that amplify the powers to reason.

Yet this history of AI should have nothing to do with the recent resurgence of covertly neo-Hegelian thought so popular with futurists and disruptive entrepreneurs, who believe in some sort of a teleological "event horizon," past which humankind will attain a posthuman singularity (all challenges met, competitors bested, obstacles overcome). A singularity that leads to the end of history sounds both contemptible and horrific—I'm not sure if humans would survive it. History cannot be personified in that way. It isn't leading us to salvation or transcendence. It has no post-human goals as such, other than those we make for ourselves.

On a less apocalyptic scale, history allows for extrapolation. It helps anticipate changes in the direction of travel. Without history, our ideals and our policies lack sufficient foresight to prepare adequately and to alter course if needed. Without history, society is doomed to struggle with long-term effects of technology, such as pollution and urban sprawl, and soon, if not already, AI.

History ultimately makes for better engineers and users of technology. Any complicated gadget contains within it a number of contingencies that appear senseless today because they were attained through a compromise with past limitations.

QWERTY, the layout of modern English keyboards, for example, was famously arranged in that order to slow down typing on typewriters that jammed otherwise. Similarly, eight bits (ones and zeroes) make a byte (a unit of information) by convention. The number eight holds little significance. The arbitrary length always confused me, until I learned that early telegraph systems also experimented with variable-bit encoding, such as the Morse code (which can have anywhere between one and five bits), as well as fixed encoding, like the five-bit Baudot code, deriving from earlier cryptographic cyphers proposed by Francis Bacon and others in the early seventeenth century. The linear ticker tape used in telegraphs subsequently could break a string of punches into any arbitrary length. It just so happens that the rectangular punch card in use by early tabulating census machines, and later by IBM, would not physically fit more than eight characters across, in rows of seventy to eighty columns stacked. The convention of an eight-bit byte (and the eighty-character code line) persisted, even though magnetic tape and later "hard drive" media were no longer constrained by those limitations. The physical properties of cardboard thus continue to inform the design of modern computers. The abstract and alien concept of "bits" became concrete to me by imagining holes punched through cardboard. On a recent trip to Kyrgyzstan, a friendly librarian gifted me a punch card used by Soviet computers in 1977. I use it as a bookmark now, but its neat rows and columns still help conjure the rather elusive "byte" as an object.

History gives meaning to such inherited structures. In following the evolution of technology, we are able to make sense of legacy technological debt. Whatever is meant by "innovation" consists of realizing which features of the inherited design remain necessary and which can be discarded or improved. Without history, the present becomes invariable, sliding into orthodoxy ("it is what it is"). History's milestones mark a path into a possible future.

It's okay, then, to sometimes struggle with basic file operations alongside some of my best data-science students. Believe in yourself, Dennis! How can I consider myself literate if I don't fully understand how writing works? A technical answer isn't enough, either. "How writing works" cannot be reduced to the present technological moment, because that moment will pass, and rapidly. Instead, writing "has come to mean" this specific arrangement of mind, body, tool, circuit, bit, gate . . . through its historical development. An obtuse technological concept, such as solid-state memory storage (the quantum shimmer behind words on screen), obtains meaning only through a specific genealogy: from ticker tape to punch card and to the NOT-AND gates flickering silently in the palm of your hand.

At the very least, the erasure of such realities stunts intellectual growth. Good engineers need history, because the progress of their education can follow the gradual development of their subject matter. History leads to deep understanding because it explains why and how something has come to be the way it is. Without it, we passively inherit a bundle of meaningless facts: it's just the way things are.

Poets steeped in engineering likewise stand to understand something vital about the nature of writing (sign, symbol, inscription, representation, if you want to get fancy). From that perspective, writing machines can be seen as the culmination of a long, mythical tradition: In the beginning, there was the letter. Then, a byte came to encode eight bits, because eight was enough to represent a character. Characters formed words and commands in a string. The string ate its own tail and learned to rewrite itself. These were the early days of autopoiesis, the writing that writes itself. What we teach today as the history of literature is also but an episode of a larger emerging storyline. Its continuing evolution must involve the people of the word.

Each turn of the historical screw brings wonder, quickly fading into the ordinary. Who remembers the ninth-century Persian mathematician Muḥammad ibn Musa al-Khwarizmi, still with us in his *al-gorithms*? What of Ramon Llull, the Majorcan monk, who studied rhetorical combination in the thirteenth century? Did he invent one the earliest chatbots with his rotating paper charts? Or do their origins lie in ancient divination charts, like the *zairajah* described by the great medieval historian Ibn Khaldun, or the *Yi Jing* (Book of Changes), written in the Western Zhou period?

Can I tell you also that Llull (pronounced *yooy*) inspired Francis Bacon and Gottfried Leibniz to propose their binary cypher systems, influenced also by the Chinese tradition? What about the *Wunderkammern* of the German baroque poets in the seventeenth century? These curious drawer cabinets—literally furniture—could be pulled in any arrangement to produce beautiful music and poetry, to the delight and horror of the audience. A long tradition of language machines also includes *An Essay Towards a Real Character, and a Philosophical Language* by John Wilkins. Written in 1668, the book imagines a new system of writing that would function as a kind of a translation protocol between cultures, commercial interests, and religions. Later, the mathematical notion of a Markov chain, central to contemporary AI, developed out of a paper on Pushkin's poetry.

These mostly forgotten artifacts remain in the background of contemporary computation. Later, we'll have the chance to pluck a few of them out—unplug and hold them for a bit to see how the whole thing fits together.

In writing this book, I was surprised to discover the history of computing permeated by distinctly literary puzzles. But to see a past shared between literature and technology has implied, also, the recognition of their altered present. Technology continues to reconfigure human orientation to the word. To study the history of literature today is to go beyond the familiar categories, like fact or fiction, or even science and the humanities, art and technology. Instead, our thread leads to an entirely separate branch of the symbolic arts, practiced on a massive scale, on the work floors of intellectual labor, across diverse industries. Here, we can find stories produced by the MITRE Corporation, one of the largest military contractors in the world, woven deep inside the Cold War–era missile defense systems; encounter fairy tales written by airplanes as part of Boeing's effort to automate their incident-reporting systems; or visit the "vector space" supporting everyday search-engine recommendations. To what genre do these texts belong? On which bookshelf should we file them? How do we read something so fundamentally unfit for human consumption?

The advance of artificial intellect obviously implicates a number of open-ended metaphysical questions. From these, I want to highlight a subset related to the subject of my expertise: literature, poetics, semiotics, philology, textual studies, the study of inscription and incantation, e.g., et al., etc.—the arcane arts.

Except not really that arcane, because though people these days read fewer things labeled "Literature," they consume more literature than ever. Text weaves through the online digital world, stitching it together. It bonds and bounds whatever is meant by artificial intelligence, in the way an encyclopedia or a library holds the sum total of human knowledge. The search engine or chatbot is just an outer garment, draped over the complication within.

A body of written works sustains all organized human activity. I don't mean to suggest that all things are textual. The world becomes truly visible to a literary scholar only through the looking glass of inscription. For example, unlike your well-adjusted physician, I know little about and am uncomfortable around bodies. A patient exists for physicians as a living body first, but that's not all a patient becomes in a hospital: Patient histories (stories) may be recorded to supplement a physician's notes. A recording of their consultation will be transcribed and transcoded into a digital file. It will be translated from one system to another, in multiple formats. It will be redacted, edited, compressed, processed, and mined for missing billable codes. This record will be sold and resold and used to train automated diagnostic bots of the future. It will be summarized, cataloged, tagged, and archived for later use. A body, in the sense of the literal Latin *corpus*, has thus been transformed into the English *corpus*: figuratively, a collection of textual records.

Much of this churn implicates automated, artificial, and otherwise intelligent work. But neither the physician nor the hospital administrator may consider it as such. Though a patient's journey from body to record leads directly to health outcomes (and to a hospital's finances), it remains without a guide, meandering through multiple, inherited legacy systems. Here, a sharp-eyed arcanist may find a worthwhile mystery: Finally, a puzzle for those who have studied exactly these sorts of things, like corpuses and taxonomies, editions and redactions, catalogs, summaries, translations, and archives—a rich trove of textual goodies!

In this book, we will similarly observe the history of artificial intelligence through the looking glass of inscription. History tells us that computers compute not only in the mathematical sense but universally. The number was incidental to the symbol. In the 1840s, Ada Lovelace, daughter of Lord Byron and one of the first "programmers" in the modern sense, imagined an engine that could manipulate any symbolic information whatsoever (not just numbers). A century later, Alan Turing extended that blueprint to imagine a "universal machine," capable of reading from and writing to an infinite string of characters. The children of Turing and Lovelace occupying our homes are therefore expert generic symbol manipulators. They are smart because they are capable of interpreting abstract variables, containing values, representing anything and everything.

Modern machines read and write. And because they carry language, we readily imagine them capable of higher-order functions like sympathy or sentience. But how do machines grow from handling inert textual matter to living language? Why do they seem to exceed their programming by telling jokes, making complex logical inferences, or writing poetry?

Metaphor holds the key to their slow awakening. Algebraic values can signify human values: $x = y$ arranges two variable entities into a familiar relation; x can also be assigned to ideas like "cat" or "dog," or an action like "call home" or "order milk." The art of manipulating symbols leads to a game of metaphors in which certain inscriptions or gestures stand for something else. Children, spies, and poets love to play games like these. When I call (after Shakespeare) the night "thick," I mean it to be particularly dark, humid, and full of anticipation. The metaphor packs a lot of meaningful baggage into a compact transport. Similarly, a machine becomes smart by "rising higher and higher through the circles of spiraling prose" (thanks, Winfried Georg Sebald!). Symbol emerges from inert hardware, gesturing toward things that we care about, like flagging offensive content, deleting spam, fighting terrorism, or indeed shopping for shoes.

Care is also a value. But can machines really soar (crawl?) from simple values like x and y to complex ones like care, equity, freedom, or justice? Reader, are you human? Is there anything inside? Can machines think, or are you just trolling me?

WE'LL GIVE SUCH TEXTUAL-PHILOSOPHICAL distractions a good go throughout the book. In conclusion, we will move to a few practical, that is to say political and economic and even personally psychological, considerations. Insert a wise, pithy aphorism about the social good, career choice, the means of production, and the future of creativity. This is the "So what?" part. How will my life change and what is to be done?

The mistake, if I may spoil the punchline, was ever to imagine intelligence in a vat of private exceptional achievement. Thinking and writing happen through time, in dialogue with a crowd. Paradoxically, we create new art by imitating and riffing off of each other. Trained on pretty much "everything ever published," AI is just another way to give that collaborative a voice. It does not, as we will see, amount to one thing but many: a talking library, a metaphor, a personification of a chorus. If we are to call it "intelligent," it is intelligent in the way of a collective. It remembers like a family does. It thinks like a state. It understands like a corporation.

Consider the self-driving car. An automobile consists of many connected subsystems: drivetrain, suspension, braking, steering, cabin control, and so on. Modern cars also contain numerous sensors, which monitor everything from speed to ambient temperature and driver alertness. Many of these subsystems exhibit intelligence in their own right. Take the humble "automatic transmission," for instance. They call it "automatic" because, unlike older tech, it automates important decisions for the driver. And it's actually quite clever! Unlike manual transmissions, most automatics won't allow drivers to take a dangerous action, like reversing mid-highway drive. (Please don't try this anyway.) The automatic will even shift to a lower gear for better braking performance when driving downhill. In bad weather, it can also send less power to the wheels, improving driver safety. Today, this unremarkable technology performs better than most human drivers.

Automatic transmissions have been around for more than a hundred years, becoming more common in the 1960s. Imagine yourself driving an automatic back then for the first time. It must have been incredible—a technology that would make driving easy, safe, and accessible. "How artificially intelligent!" you would've said.

Nobody said that, I know. Whatever magic there was, dissipated. Just a while ago, you may remember having been impressed with your automobile's global positioning system (GPS). We drive by the stars! Or how about rear-view cameras? And self-parking! Eventually, car-makers hope to arrive at something like "Level 5: 'Full Driving Auto-mation,'" involving "sustained and unconditional [. . .] performance [. . .] without any expectation that a user will need to intervene" (as specified in J3016-JUN2018, *Surface Vehicle Recommended Practice*, by the Society of Automotive Engineers). Yesterday's remarkable automa-tions have been left behind at Level 1, "driver assistance," limited to specific "operational design domains," alongside cruise control, adap-tive suspension, and automatic windshield wipers.

As the examples above illustrate, intelligence evolves on a spectrum, ranging from "partial assistance" to "full automation." Many everyday tasks simpler than city driving have been automated somewhere along that spectrum. Your average kitchen toaster, for instance, requires no input except bread. A user intervenes only to start the device and to consume its delicious outputs. Granted, the procedure required for the task—an algorithm—wasn't very complex to begin with: lower bread, fire, eject. By automotive standards, our toasters have probably achieved Level 5 automation already, or at least Level 4, where all toasting tasks are automated under specific circumstances of the kitchen, with human override for an occasional bread malfunction. Hopefully soon they will be on their way to greater achievements, like spamming our phones with photos of burnt toast.

I don't know much about cars or toasters, to be honest. Rather, I'm interested in the appearance of a small imaginary human from behind the curtain. Why, for example, does an automatic transmission seem like a boring tool, whereas a drug-discovery algorithm was recently described by Henry Kissinger as a thing able to "master its subject," "devise new strategies," and even "detect aspects of reality humans have not detected, or perhaps cannot detect." How does a mere tool move into the subject position of the sentence—where it detects, devises, and masters—gaining a sense of agency and interiority in the process?

Plenty of smart things don't get to do that. Other less-intelligent ones readily conjure the image. Savvy marketeers understand that self-driving cars and smart houses make for compelling subjects, while automatic transmissions and toasters do not. Emergence somehow happens by the critical accumulation of smarts, where intelligence sometimes bubbles up, pulling along with it the more intangible silk of sentience, awareness, and conscience.

Leaving the world of culinary and automotive hardware, I return to the bundle of specifically textual technologies, such as chatbots, machine translators, named entity extractors, automatic text summarizers, speech recognizers, spam filters, search engines, spell-checkers, classifiers, auto-taggers, sentence autocompletion, and story generators. (Let's not forget, too, that whatever is meant by "AI" includes other, non-verbalizing tech as well, from weather-prediction models to drug-discovery algorithms, robotics, and image classifiers.)

Robots that chat impress us in a particularly satisfying way, possibly because language always hints at a worldview (a mind?) within. Modern chatbots don't contain a mind as such. Or, rather, they express minds metaphorically, through a fuzzy, loose analogy with human brains. Yet, without knowing much about the world at all, they are capable of synthesizing information, responding to questions, and even writing college-level essays.

How did this sleight of hand come to be and why? Once we see intelligence on a spectrum, integrated within almost every domain of human life, we can begin to unwind the metaphor of its collective "mental" development, by stages and significant milestones. In the domain of language, we can draw a line between fully automated text generators back toward familiar tools like spell-checking and word autocompletion, and further to long-standing historical practices and algorithms.

Sufficiently inspired by these preliminaries, I am now ready to discuss the historical frays of textual technology proper. Don't get hung up on the imagery of pioneers or milestones, by the way. Who did what first or last is irrelevant and usually misguided. Al Gore did not invent the internet. Ideas come into being as physical things with great effort, through multiple parallel iterations. For this reason, the several lengthy strands I picked out for you in the next chapters are merely representative of a trend. Each draws its own particular trajectory—a magical incantation—which we can retrace and recite together.

CHAPTER 2

Letter Magic

■ ■ ■ ■ ■ ■ ■ ■

WHEEL

Let me let you in on a little secret: *koldun*, the word meaning "sorcerer" in several Slavic languages, arguably originates from the name of one of the most prominent medieval (fourteenth-century) scholars, Abd ar-Raḥman ibn Muḥammad ibn Khaldun al-Hadhrami—the son of eternity from Hadhram—or simply Ibn Khaldun. In his epic history of the world, modestly titled, *Muqaddimah* or "Introduction," Ibn Khaldun documented the use of *zairajah*—"a remarkable technical procedure [. . .] for alleged discovery of the supernatural":

The form of the *zairajah* they use is a large circle that encloses other concentric circles for the spheres, the elements, the created things, the *spiritualia*, as well as other types of beings and sciences. Each circle is divided into sections, the areas of which represent the signs of the zodiac, or the elements, or other things. The lines dividing each section run to the center. They are called chords. Along each chord, there are sets of letters that have conventional (numerical) value [. . .] Inside the *zairajah*, between the circles, are found the names of the sciences and of topics of the created (world). On the back of (the page containing) the circles, there's a table with many squares, fifty-five horizontally and one hundred and thirty-one vertically. Some of the squares are filled in, partly with numbers, and partly with letters. Others are empty.

Ibn Khaldun further described several such devices, giving us a glimpse into their "remarkable techniques." In short, a learned sooth-sayer would write down a question, converting its letters into numbers. These would then be transposed back into letters, by consulting a number of intricate charts, according to "well-known rules" and "famil-iar procedures." Several such computational "cycles" would produce a shortened string of letters, which could finally be expanded into a rhymed answer. With proper training, the zairajah could obtain the "knowledge of the unknown, from the known," Ibn Khaldun explained, giving the intellect an "added power of analogical reasoning."

Despite his enthusiasm for the device, Ibn Khaldun also under-stood its limitations. The zairajah revealed analogies implicit in the question itself, through a relationship between an "arrangement of let-ters" and another "arrangement of letters." No amount of verbal com-bination could attain a "conformity of the words to the outside (world)." "We have seen many distinguished people jump (at the opportunity for) supernatural discoveries," Ibn Khaldun wrote. "They think that correspondence (in form) between question and answer shows corre-spondence in actuality. This is not correct." The answers rather revealed a logic implicit in the language itself. Truth value lay outside of lan-guage. The wheel's answers "remained veiled," requiring further verification. "One should not think that one can get to the secret of the letters with the help of logical reasoning," Ibn Khaldun insisted. "One gets to it with the help of vision and divine aid." Modern, algorithmic soothsaying contains the same defect. Any artificial language system may at any moment lose its grasp of the real world and begin hallucinating or fabricating imaginary facts.

For his own illustrative purposes—to give you an idea of fourteenth-century AI performance—Ibn Khaldun used the zairajah to pose a self-referential question (typical of the genre today, also). "We wanted to know," he asked, "whether the zairajah was a modern or an ancient science." The question was then broken down into ninety-three chords, computed through twelve cycles on the wheel, in a procedure that took the author several manuscript pages to describe in detail, producing the following rhymed answer: "The Holy Spirit will depart, its secret having been brought forth / to [the prophet] Idris, and through it, he ascended the highest summit."

What are we, modern online shoppers, to make of ancient "letter magic," as Ibn Khaldun called it? A few ideas come to mind.

First, pretty impressive! Though difficult to follow, the method produced an intelligible and intelligent answer. One suspects the involvement of some cherry-picking and hand-waving, but so is the case with AI today, seven centuries later. From ELIZA—the famous chatbot therapist meant to prove how easy it was to fool humans—to *The Policeman's Beard Is Half Constructed*—a collection of algorithmic poetry that seemed groundbreaking at the time it was published in the 1980s—scholars have often plucked the few clever bits from reams of incomprehensible logorrhea.

Second, whoa, these things are old! Zairajah exemplifies a number of divination techniques extant in the medieval Muslim world, but also found in antiquity, going back to Greek, Chinese, Egyptian, and Sumerian roots. Text generators turn out to be as old as written language itself.

Third—okay, wading into the weeds now—from its earliest documented use, letter magic promises to reveal a logic inherent in language itself, seeming like a shortcut to hard-earned scientific knowledge (vision and unveiling, in medieval scholarly terms). And because the various people of the letter considered their textual traditions sacred, procedural word manipulation was always proximate to magic, if not outright divinity.

Modern AI systems are often discussed similarly in the context of ineffable mysteries such as consciousness, creativity, or, Khaldun-forbid, the singularity (an irreversible event horizon at which AI transcends its human origins to become Nietzschean superintelligence). Then as now, among the philosophers and entrepreneurs eager to jump on the bandwagon of AI, we find plentiful conmen and charlatans.

Finally, note that even the most advanced modern, language-based AI systems today continue to have a problem with external validation—an issue already well understood by Ibn Khaldun and his contemporaries. Procedurally generated text can make grammatical sense, but might not always make *sense* sense. Noam Chomsky, the MIT linguist, famously expressed this paradox in this sentence: "Colorless green ideas sleep furiously." Derived from Chomsky's procedural grammar, the sentence may be correct according to the rules of language. But to know that ideas don't sleep, or that one does not sleep furiously, requires experience in the physical world.

Letter magic is still practiced in the prison of language. No matter how complex the operations, words generate only more words. Their truth or falsity obtains through lived experience. This significant limitation precludes the idea of something like "ethical AI," because language itself contains within it all possible values, good and bad. External constraints are necessary to compel it toward goals that exist (and evolve) outside language, in the world that includes, you know, all the other stuff, like guns and germs and capital and steel. Combinatorial language games generate sentences without limit. Ethics require limitations like pain, illness, loss, and death.

TABLE

So far, we've managed to dip our toes (fall off a cliff?) into the magical waters of soothsaying and divination, in considering the medieval Arabic Q&A bot—a set of "remarkable techniques" for language manipulation.

A careful examination of the zairajah and similar devices reveals several distinct attributes, each worthy of its own little genealogy.

Note, for example, that despite its circular shape, the divination wheel can also be seen as a type of table or chart. Charts aren't as simple as they appear. Unlike the messiness of presented information, on any topic, a chart boils things down to their essentials. Charts are therefore by nature more compact and more portable than the source of their original, non-processed data. This property will become important later, when we consider their capacity to encapsulate and transfer intelligence removed from its origins.

At the very least, a table presents information in a systematic way, usually containing rows and columns. A chart implies order, however simple. Here's my fake shopping list, for example:

Product	Price	Quantity	Type
Oatmeal	$4.99	1 package	Steel-cut
Chocolate	$3.48	3 bars	Dark
Blueberries	$9.75	2 bags	Frozen
Milk	$3.15	½ gallon	Whole

Aside from being impressed with my organizational skills and passion for antioxidants, an alien visiting our planet for the first time would learn that foodstuffs fall under the category of "products," and that products contain such attributes as "price," "quantity," and "type." A table therefore already implies a sense of hierarchical labels, or a taxonomy. Moreover, the taxonomy is "controlled," in the sense of containing a limited number of allowed values. Of course, I am not so formal with my shopping lists. You would be surprised, however, to see milk requested in units of "cow" instead of "gallons," for instance, or blueberries of the type "decorative." By limiting the available language choices, a taxonomy attempts to capture the relationship between things in the world, where products have prices and chocolate is sold by the bar.

Taxonomies of all types were synonymous with medieval science, as well as being an important topic in contemporary logic, medicine, law, and computer science. Consider a field like medicine, where the infinity of analog bodily sensations must be distilled into a controlled set of named and billable conditions—a diagnosis. The American Psychiatric Association publishes the infamous *Diagnostic and Statistical Manual of Mental Disorders*, now in its fifth edition (*DSM-5*). Your insurance is unlikely to cover a health issue not named in the manual. Conversely, aspects of the human condition now considered in the range of normal (neurosis, homosexuality) were once unjustly included, to the detriment of the people they labeled. A part of ancient arcane arts, taxonomies still rule the world. Think of electronic databases working in the background of every hospital as giant zairajah circles, spinning a yarn that connects patients, physicians, pharmacies, and insurance companies.

The greatest popularizer of divination circles in the medieval Western world was Ramon Llull, who, in the wake of the Christian reconquest of Muslim Spain, both borrowed and extended the system to create his "new logic." The Art, as he called it, made extensive use of rotating spheres that could, by the use of a complex notation, represent knowledge about anything and everything. In addition to being portable and categorical, Llulian logic was symbolic, rule-based, and combinatorial.

Let's take these individual properties slowly and by turn.

Start by examining the following partial table transcribed from *Ars Brevis*,

	1st Wheel	2nd Wheel	Questions	Subjects	Virtues	Vices
B	goodness	difference	whether?	God	justice	avarice
C	greatness	concordance	what?	angel	prudence	gluttony
E	power	beginning	why?	man	temperance	pride
H	virtue	majority	when?	vegetative	charity	ire

Note that, initially, the symbols BCEH in the first column could be "overloaded" or have multiple meanings, symbolizing questions, subjects, virtues, or vices. For instance, B could, depending on its position, stand for "goodness," "difference," "whether?", "God," "justice," or "avarice." Llull explained that:

we have employed an alphabet in this Art so that it can be used to make figures, as well as to mix principles and rules for the purposes of investigating the truth. For, as a result of any one letter having many meanings, the intellect becomes more general in its reception of the things signified, as well as in acquiring knowledge.

Pragmatically, the overloading of the lettered variables made the charts more universal. Had they contained only a single meaning, they could have been used for a single purpose only. But Llull's art worked equally well for theology as for medicine.

Further, the arrangement of letters in figures rigidified the procedure of composition. The shape of a wheel or a tree, common in medieval documents, made only certain of the combinations possible. Where Ibn Khaldun documented pages upon pages of complicated procedure, Llull crafted a visual mechanism, consisting of two nested wheels, rotating freely. In use, the mechanism fixed the range of possible symbolic combinations to those that produce only the valid results. "Everything that exists," Llull explained, "is implicit in the principles of this figure, for everything is either good or great, etc., as God and angels, which are good, great, etc. Therefore, whatever exists is reducible to above-mentioned principles." In other words, rotating the wheel would align entity "angel" with attribute "prudent" or "just" but never "evil," generating all true combinations, and never the false ones.

Finally, the features above enabled a combinatorial process, by which all the possible inputs could be combined to generate all possible outputs. These, Llull documented in another "table of tables," containing values like BDTC, CETC, EGTF, FTEG, HITG, and so on. By such means, the system was meant to complete gaps in existing knowl-

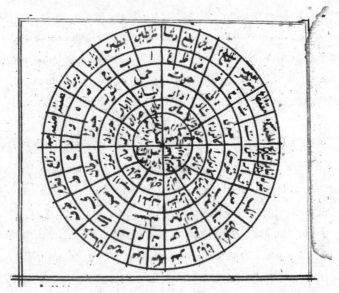

Zairajah—medieval Arabic divination circle used for astrology by Ahmad al-Buni as described by ibn-Khaldun in his Muqaddimah *(Introduction). Ahmad ibn ʿAlī al-Būnī,* Shams al-maʿārif al-kubrá *(1874), 17.*

General Collection, Beinecke Rare Book and Manuscript Library, Yale University.

edge: If God is Great, God is also Good and Powerful and Virtuous. And if you didn't realize a few of those before, now you did.

Heady stuff. Imagine how impressive and exciting Llullian complications must have seemed to the learned folk of the time! Particularly if you were the average sort of a monkish scribe, somewhere in young Spain or Germany. You never did learn to read or write that well. Your job was to transcribe manuscripts in the local parish. You did so mechanically, without really following the complicated theological argument at hand.

Until—what's this?—a neat structure interrupts the dense wall of Latin text. "This table is the subject, in which the intellect becomes universal," you read. The name of Llull came across your carrel before.

You heard stories of his study of Arabic and his ministry among Andalusian Muslims. "By said table, it [the intellect] understands and abstracts many particulars objectively," the text reads, "and this it does by applying to each question one argument being extracted from each compartment of a given column." Here, you manage to glimpse a pathway into the inner sanctum—a step-by-step method for discovering truth. "The intellect banishes doubt, remaining in it calmly and positively," you read along with Llull. "It now knows itself to be completely general and artful, and clothed in great knowledge."

Calm, positive, and clothed so, I am thoroughly exhausted of possibilities. Before we are lulled to sleep, however, notice the transpiring sleight of hand. Let's follow the original Latin, instead of the usual published translation.

How did the intellect come to know itself? What does that even mean? The table (*tabula*) is the subject, to repeat after Llull, in which the intellect fashions itself (*se facit*) as a universal, by understanding (*intelligit*) and abstracting from the particulars. What a strange way of putting things! The table captures a disembodied intellect. There, from each cell (*camera*) of a column, it derives (*abstrahitur*) a calculation (*una ratio*)—or perhaps, the English ratio, in the sense of a "quantitative relation between amounts." It's not clear whether Llull means to give us commonplace instructions (something like "Consult the letter combination in each cell"), or to say something profound about reason, in the sense of "Reason attains intelligence inside the table."

What just happened then? Did we really witness the emergence of artificial intellect—"an intellect that fashions itself"? Not exactly. The sleight of hand happened through personification. Clever Llull fashioned a clever system—clever in a manner of speaking. Yet it continued to operate cleverly in the absence of its maker. So, it did contain some smarts after all! These were left behind by Llull, who's long gone. But were you to give me a clever explanation, we wouldn't call the explanation itself smart, right? Your intellect would be the source of intelligence, transmitted via the explanation. Yet, in common usage, cleverness tends to detach from the human, almost as if to appear self-sufficient.

A wizard has slipped behind the curtain. To help in the translation of Latin, I used the Latin Word Study Tool from the Perseus Project, which compiles more than sixteen million Latin words, presenting a complete concordance of word meaning, in use. Did it actually "compile" or "present" anything, or was that done by Gregory Crane and his team from Tufts University, the makers of the project? Their smarts made me smarter. But how is it that we find it so natural to cede agency to the tool?

At what point, on the scale of automation, from partial assistance to independent reason, does the intellect begin figuring in the grammatical subject position? And why is it always the one intellect or intelligence generally—an indefinite, singular noun? I cannot wrap my mind around that. Llullian diagrams and digital Latin study tools contain many different components: databases, charts, and wheels. Does their intelligence emanate from one or many sources? Or, does it hover like a cloud above our collective minds and Latin Word Study Tools?

Practically, the technology itself begins to exceed the rather limited cognitive capacity of individual humans. Before anyone takes offense, understand that together we are an impressive bunch. But individually, I am at least aware of my own substantial limitations in driving, reckoning numbers, or arranging letters. Automated assistants, like driving aids, calculators, or spell-checkers, offer a measure of welcome performance enhancement. I know comparatively little about the world by memory. But surrounded by dictionaries, encyclopedias, and search engines, I know a lot—though my actual share of that knowledge remains minuscule.

When encountering a smart gadget, I'm therefore impressed not so much by its local effects on me, personally, but by its capacity for shared, collective achievement. In writing with a spell-checker enabled, I am writing with a shadow team of scholars and engineers, able to extend their aid across time and space by the virtue of technology. This collective endeavor resists easy description. I am not eager to cede authorship or to continually credit a school of distant collaborators. I wrote this, after all. Yet I also must admit to a diminished share of authorial agency, in this and many other matters of intelligence.

Spell-checkers and dictionaries have a long history. I don't normally have to think about them, however. Instead, we employ a cognitive-linguistic shortcut by condensing and ascribing agency to the technology itself. The wheel divines, not Llull or Ibn Khaldun. And it's easier to say "*The phone* completes my messages" instead of "*The engineering team behind the autocompletion tool writing software based on the following dozen research papers* completes my messages."

This strategy works, until the metaphor threatens to run away with unintended meaning. The smart table, in the words of Karl Marx, another furniture enthusiast, "changes into something transcendent," a fetish, turned onto its head, fashioning grotesque ideas out of its wooden brain, as it recedes into the backdrop of ordinary commodities. "I made this," I keep saying to myself, but when the tool breaks, I curse it and call it a "stupid thing."

CHAPTER 3

Smart Cabinets

■ ■ ■ ■ ■ ■ ■ ■

T WO GERMAN BAROQUE POETS WALKED into a bar. Well, more like, one was a poet and the other a scholar. And they corresponded by letter. Athanasius Kircher, a neat man in his forties, spoke and dressed plainly. An established figure by the time of their virtual meeting, he was slated to take Johannes Kepler's place at Habsburg court as the royal mathematician. In years past, he had taught mathematics and ancient languages at Wurzburg. Fate found him at the Collegio Romano in Italy, where his research program expanded to cover geology, chemistry, and biology.

The eager attention from a flamboyant younger admirer clearly annoyed him. The gaunt Quirinus Kuhlmann struck an odd figure. A childhood speech impediment left him to pursue a career in letters with fervor. At the University of Jena, known as a bit of a party school in the early sixteen hundreds, he thought little of his peers or teachers. Instead of studying law as planned, he cultivated an image of a brooding poet, claiming to receive divine visions through illness and hallucination. Rumors of heresy circulated, as well as of his "colossal egotism."

At Jena, Kuhlmann's scholarly brooding included the study of Llull's cryptics, as they appeared in Kircher's work. By this time, and despite its own brush with the Inquisition, Llull's Art had spawned numerous adherents across Europe. The emerging science of combinatorics promised shortcuts to divine revelation. Under the spell of Llull, Kuhlmann wrote a series of unprecedented technical sonnets, among them "Heavenly Love-Kiss XLI," meant to showcase his understanding of combinatorial composition. I'll translate a snippet for you here:

> *From Night / Haze / Fight / Frost / Wind /*
> *[. . .] / Fire and Plague*
> *Follows Dawn / Blaze / Blood / Snow / Still /*
> *[. . .] / Blight and Need*
> *From Loath / Pain / Shame / Fear / War / Ache /*
> *[. . .] / Jest in Scorn*
> *Wills Joy / Pride / Fame / Trust / Wealth / Rate /*
> *[. . .] / Steady Days*
> *[. . .]*
> *All changes, all loves, all seems to hate something:*
> *All earthly wisdom comes to those who reflect thusly.*

The sonnet's concluding lines invite the reader to rearrange the words at will—a radical gesture. The many resulting combinations produce individual meaning, amounting to the sum total of human knowledge, Kuhlmann explained. Despite such grandiose claims, Kuhlmann abandoned the technique in other compositions, making me think he saw this particular "love-kiss" as a technical experiment, not much more.

At stake in the virtual meeting between poet and scientist was

Kircher's recent invention called the Mathematical Organ. Waiting for their ale, both men contemplated the device in front of them. Made of polished, "artfully painted" wood, the Organ resembled a large box. Opening its lid revealed a row of labeled wooden slates, filed vertically: four columns of fifteen narrow slates, four columns of seven wider ones, and four columns of five of the largest pieces. When pulled out, each of the slates contained a string of letters. By consulting the included booklets—which he called, wait for it, *applications*—any combination of the planks cohered into a complete, harmonious composition. Cleverly, depending on which manual was consulted, the same organ could be used to compose music, write poetry or secret messages, and even do advanced math, such as reckoning the Easter calendar.

The conversation grew more heated with ale. Kircher took the same systematic approach to all his research projects. The box was made according to the latest scientific principles. He had recently sold a version of the device to the young Archduke Charles Joseph of Austria. The child would use it in his studies, with the help of a trusted tutor. Such devices, Kircher argued, were easy, delightful, and instructive. They were, above all, useful, allowing untrained operators to create and to perform. Technology was there to serve, making difficult knowledge accessible to a larger audience.

Kuhlmann objected. The path to knowledge should remain tortuous, accessible only to those (like him!) willing to walk it properly. The box is just an ingenious game, my ingenious Kircher! Kuhlmann was starting to slur his words a little. The intelligence lies with you. Without the box, the young duke remains an idiotic parrot (I'm sorry, he really did write that, only in Latin). Nothing is retained on the inside. And without inward understanding, the child captures neither knowledge nor intelligence.

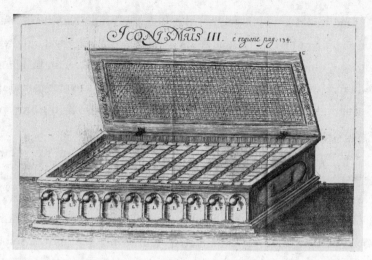

A part of the Mathematical Organ, as described by one of Kircher's students, P. Gaspar Schott. Printed by Johann Andreas Endter in Würzburg (1668), 135.

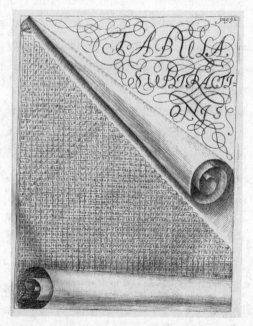

One of the "application" tables included with the device. Schott (1668), 93.

One would hope, in another timeline, the young Kuhlmann was not planning to travel Europe to say such things about royal children out loud, especially not in the presence of, let's say, a certain fledgling and repressive Eastern regime. Though his talents were recognized in Germany, he would later burn at the stake in Moscow, by the order of Ivan V, denounced by his fellow Lutherans for iconoclasm.

We arrive then, as always, at the ghost in the shell—of smart furniture. Clearly, neither poet nor scientist thought of the Mathematical Organ as an animated entity. They differed rather in the idea of the human involved. For Kuhlmann, the tooling edge of combinatorial poetry aimed inward, toward the reconfiguration of the spirit. Combinatorial sonnets helped precipitate a mercurial spiritual transformation, difficult for us to recreate sober and in my translation. For example, the sonic parallelism between *Haze* and *Blaze* (which I mirrored from the German), encouraged readers to seek further, conceptual similarities between smoke and fire. Name and physical property momentarily aligned. New, unexpected connections were meant to be made through private internal reconfiguration.

By contrast, Kircher's machine con-figured externally, producing a specified rational effect—be it the calculation of numbers or composition of poetry—which also happened to be useful, elegant, or delightful—all outward qualities.

The two views on the machine tell us something important about intelligence, which manifests both as a private experience of inward understanding and its outward, instrumental effects on the world. Let's take a moment to consider both. On the former view, we'll call it Platonic, intelligence means something like "the correct internal alignment of thoughts and feelings with universal truth." On the latter, we can call it Aristotelian, it implies "a universal capacity to achieve specific results."

Notice something strange?

In the Platonic model, intelligence works as the SOURCE of action. It therefore necessarily exhibits qualities of being private and local, ineffable even. Therefore, paradoxically, Platonic universal rationalism tolerates contingent, nonrational paths to enlightenment. Of course! I think to myself. There is no one general way of being smart. How would that work? Sometimes I can't even articulate how I've come to understand certain things. It just happens or doesn't. In that sense, our intelligence is bound by our individual bodies, education, and personal history. It therefore resists one single universal description.

Private experience doesn't have to be wholly subjective or arbitrary. Outside access tends to be difficult, however. How can a teacher glimpse a student's view from the inside, so as to better ascertain whether they've learned something properly? Plato—well actually, the fictional character called Socrates—famously mistrusted all writing in general and rote memorization in particular (despite being a memorable writer). An often-cited dialogue recalls Socrates encountering Phaedrus, a student of his, who was on the way back from another lecture. When Socrates asked him about the content of that lecture, Phaedrus retold the words from memory. Socrates suspected a parrot, however, and proceeded to question Phaedrus poignantly, exposing the limits of his student's understanding. Similarly, any fixed, mechanized, external trappings of acumen risk being disconnected from the source. Instead of smarts, you get parrots, who, also according to Descartes in his *Mediations*, merely repeat without understanding.

There can be no shortcuts on the road to knowledge. A student achieves internal change through mental toil. The Platonic is therefore necessarily also a human-centric view, because, on the inside, I don't know how to be a bat or a rock or a smart table or anything else. Platonic intelligence implies appropriate personal transformation.

In the Aristotelian model, intelligence is the GOAL of thought. It doesn't matter how it is achieved at the source. One jumps into the collective pool called "intelligence" through appropriate action. And all sorts of stuff besides humans float in that pool: bats, tools, automatic transmissions, smart tables, mathematical organs, smart phones, etc. This makes sense to me also, because when I think about stuff, I usually don't do it on the inside only: I pace, take notes, read books, and talk to friends. Looking things up on Wikipedia, consulting a table, following a certain procedure I learned from a textbook, or evoking a "smart assistant"—none of these detract from intelligence in the Aristotelian sense.

It would be foolish in fact to refuse such outside assistance. I cannot imagine knowledge that develops in complete isolation from the world. (Incidentally, the twelfth-century Andalusian treatise on the "Self-Taught Philosopher," titled *Hayy ibn Yaqzan* in the original Arabic by Ibn Tufail, was one such attempt, prefiguring the modern "castaway" genre, represented by Daniel Defoe's *Robinson Crusoe* (1719) or television series like *Lost* (2004–2010) or the film *Cast Away* (2000), starring Tom Hanks.) Science itself would be impossible in isolation because it produces intelligence in a way that's fundamentally social and collective. By developing novel tools and techniques for externalized thought, like Aristotle or Kircher, a scientist contributes to the overall shared capacity for reason. In that sense, something becomes right or true by its measure of accordance to an established body of work. Intelligence in the Aristotelian sense is subject to external review. We don't have to concern ourselves with "what's really going on, on the inside." If it shimmies and quacks like a smart duck, it's a smart duck.

That the two senses of *intelligence*—Platonic and Aristotelian, internal and external, private and public—both use the same word causes much confusion. It would be simpler if we had at least two distinct words for the Platonic and the Aristotelian definitions, capturing the difference between intelligence as the SOURCE of and the GOAL of action.

Baroque birds of all sorts bring us a step closer to the story of modern digital computers, able to mimic any other "discrete-state machine" universally. Called a "universal machine" by Alan Turing, the computer promises to model general intelligence, regardless of the task it was originally designed to perform. Recall the obsolescence of single-purpose machines like calculators, cash registers, and digital music players. The universal computer ate them all. Similarly, general intelligence refers to the ability of an individual or machine to learn, reason, and problem-solve across a wide range of domains and contexts. In this sense, general intelligence is often presented to us in the guise of "universal intelligence," intelligence capable of performing any cognitive task.

Given the discussion above, the words *general* and *universal* should be treated with suspicion. Universalism, for all its strengths, comes with a host of problems, already apparent in our case studies.

From the Platonic, internalist perspective, how smarts are achieved matters. I might value good grades on an exam, for example, but not if they were the result of cheating. Conversely, I don't usually care how my refrigerator thinks, only that it does its job. Anything goes for a smart machine, evaluated on efficacy alone. A machine may perform intelligence by any obscure means. My students must show their work.

The "private" ways of machine intelligence also may not necessarily agree with my own, or that of my family or community. Efficacy sometimes comes at a cost. For instance, it may be that my television gets smarter by continually recording conversations in my living room, so as to better tailor its recommendations to my family. These means do not justify the ends for me. The so-called alignment problem stems from an impassible paradox: generally universal intelligence does not always align with local values. Smart in the general sense might not be so smart in the specific context. We should therefore always be cautious with claims for intelligence in the general, universal sense.

From the Aristotelian, externalist perspective, the idea of an intelligence considered "universal" can be equated to "generally applicable." General application contains no impassable paradoxes, with one huge caveat: where it gains in universal application, it loses on smarts.

Tasks that can be automated become devalued. Witness numerous smart devices, like toasters and thermostats, that modern humans treat with disdain. While inspiring awe initially, smart technology of the past becomes mindlessly assistive, no matter the expanded ingenuity. The same can be said for the tools of intellectual trade. Dictionaries, grammars, thesauruses, and encyclopedias were once hailed as monumental national achievements. Today, they are silently integrated into digital autocompletion or autocorrection tools. We don't even think of them as sophisticated technology.

As our tooling improves, the functional definition of intelligent action changes as well. Because the externalist perspective is grounded in the world, general intelligence recedes beneath the rising tide of average achievement. For instance, at one time, the mere fact of being literate would have been considered exceptional. One could build a career of simply reading and interpreting texts for an illiterate public. With mass literacy, skills of reading and writing migrate from the upper to the lower bounds of intellectual work. Similarly, while the ability to memorize complicated calendar patterns may have been impressive a few centuries ago, spending time on unaided astronomical reckoning today would be considered foolish, with the most advanced astronomers being those able to use assistive technology (like powerful computing) most effectively. In the applied world, intelligence bubbles above shared capacity. But no matter how cleverly effervescent, once a smart tool reaches general adoption, it dissipates into the surface of baseline intelligence.

Here we reach the paradox of the functional idea: Exceptional performance cannot become universal, by definition. A "smart" device is merely smarter than the previous generation. Once adopted widely, it passes into the average.

UNIVERSAL CHARACTER

The attempt to overcome the said limitations of universal reason is indicative of computing history in general: from divination wheels to baroque composition cabinets and Turing universal computation machines.

There was, in the seventeenth century, a great deal of interest paid to universal languages in Europe, for reasons beyond poetry or philosophy. The Protestant revolution placed a premium on vernacular languages, displacing Latin from its privileged, liturgical position. Bibles had to be read and translated locally. Issues of what should count as the "correct" translation could and did grow into armed conflict. Printed books, like those by John Wilkins, were becoming more commonplace in the wake of the Gutenberg printing press. Issues of universal literacy now seemed pressing. The establishment of the Royal Society in London, among other learned societies, made publishing—alongside a clear writing style, peer review, and discussion—an integral part of the scientific process.

As Kircher's experiments with the Mathematical Organ have

shown, reason could become subject to automation. Yet how could we verify the results of any machine that claims to produce knowledge in excess of our own intellectual capacity? What if the smarts generated by a mechanism turned out to be nonsense? We would have to become smarter than the machine to check its work. We could do this, for example, by validating generated results against an established sense of the "ground truth." A machine capable of arriving at established truth independently could be said to have cleared the bar of artificial reason. But what about claims to entirely novel ideas? What if machines were to discover new answers, breaching established grounds? How could we tell such "truths" apart from mechanical gibberish?

Enter universal language machines.

The principle of Kircher's cabinet lay in the included reference tables, governing the rules of composition, a grammar. This class of machines was prone to producing nonsensical answers—the colorless green ideas sleeping furiously—despite sophisticated programming. Skeptics like Kuhlmann pointed to the arbitrary nature of the sign as the culprit. Because of their general imprecision, natural human languages yielded poor building blocks for mechanized reckoning. Words arranged according to predetermined rules did not always correspond to the proper arrangement of things in the world. Language and physics played by different rulesets. Grammar and sense did not always align.

Wouldn't it be wonderful if they did? What if Kuhlmann's chance resonances—*Haze/Blaze, Shame/Fame*—could lead to actual scientific discoveries? If machine inputs could be formalized to contain pure, unambiguous truths, arranged according to inarguably correct principles, the outputs were bound to be true! The alignment of sense and grammar seemed to be a promising direction for new technology.

In the 1660s, John Wilkins, an Anglican polymath and one of the founders of the Royal Society of London, published just such a proposal, titled *An Essay Towards a Real Character, and a Philosophical Language*. In this monumental, five-hundred-plus-page tome, Wilkins proposed an artificial language, incapable of falsity. In it, the internal and the external would become one. The rules of grammar and the rules of physics could join in harmony, providing the ultimate tool for clear reasoning. (Or so he thought. The book suffered a cool reception.)

Science, bound to become the language of universal truth, he reasoned, should base its communication on grounds more solid than natural language. Consider the mischief done, he wrote, "and the many impostures and cheats that are put upon men, under the guise of affected insignificant phrases." Pseudoscientific "charlatans" make their reputation by using "pretended, mysterious, and profound notions, expressed in the great swelling of words," leaving the "internal notion of apprehension of things" confused with the "external impression of these mental notions." "So that if men should generally consent upon the same way or manner of *Expression*, as they do agree in the same *Notion*," through the use of Real Character, "we should then be freed from that Curse in the Confusion of the Tongues, with all the unhappy consequences," Wilkins wrote.

To these awesome ends, he proposed—you guessed it—the smartest of all tables. Step One, the world. The first thing to be done, according to Wilkins, was, modestly, "the enumeration of and description of such things and notions as are to have *Marks* or *Names* assigned to them." Wilkins must have realized the enormity of the task, admitting that perhaps his giant classification schema contained a few "redundancies" and "deficiencies." Still, in the course of the first three-hundred-some-odd pages, he braved a grand "scheme" that captured the state-of-the-Royal-art. This frankly madcap (even by its own, contemporary standards) project enumerated anything and everything under the sun: "transcendental relations mixed and general"; "herbs according to their leaves," shrubs, trees, and grasses; stone and metal; fish, bird, and beast; magnitude, space, and measure; a catalog of habits, manners, and diseases; relations civic, judicial, military, and ecclesiastical; among many, many other "notions that fall under discourse."

Having listed all possible named things known to science, Wilkins moved to Step Two, in consideration of a "philosophical grammar," containing the rules of their possible recomposition. Here, the grammar of the internal, conceptual world could be brought to accord with natural law, governing the morphology of the physical world. Though unclear as to how this would work in practice, it was apparent that Wilkins imagined grammar, in its usual sense, to be coupled to something like the grammar of physics or chemistry, making language and physical world function under the same rules. A false idea in this new language would be, if not impossible, then difficult to articulate. Words would work like Lego blocks, snapping together only in ways that made sense.

This was also the weakest part of his system because the accordance of the two grammars depended on his initial haphazard classifications.

In the third and concluding step, Wilkins proposed a novel writing system in an attempt to solidify the relationship between word and thing, thus breaking the spell of arbitrary signification. Words constructed according to "Real Character" fit together like parts of a device, which, when activated, produced only true sentences. "Now, if these *Marks* can be so contrived," he wrote, "as to have such a *dependence* upon, and relation to, one another, as might be suitable to the nature of the things and notions which they represented [. . .] the *Names* of

The Lords Prayer.

1	2	3	4 5	6	7	8	9	10	11

Our Parent who art in Heaven, Thy Name be Hallowed, Thy

The Lord's Prayer rendered into universal character by John Wilkins. John Wilkins, An Essay towards a Real Character and a Philosophical Language *(1668), 395.*

things could be so ordered, as to contain such an *affinity* or *opposition* in their letters and sounds, as might be some way answerable to the nature of the things they signified."

The alphabet, as the author admitted after transcribing the Lord's Prayer into his system, might not have turned out as practical as he had planned. Still, he hoped it delivered a distinct advantage over European alphabets, mistakenly believing Chinese or Arabic writing systems to be somehow more intrinsically closer to his ideals. The Real Character would render each word precisely, according to the classification of things and relationships located in the main table. And though Wilkins could not achieve a perfect analogy between word and thing, he thought that further improvements of his system would someday enable great advances in scientific communication, commerce, and even facilitate peace between nations.

Judging by the many wars and disagreements since, this experiment failed. Still, Wilkins sowed yet another seed of universal language. And it would bear fruit in the universal machines made by Leibniz, Babbage, Lovelace, and, later, Turing.

STEPPED RATIOCINATOR

In 1679, Gottfried Wilhelm Leibniz wore a frizzy wig. Like other great men, he was modest about his accomplishments. "I'm not sure why Aristotle or Descartes didn't consider this amazing thing I did while still a boy," was a common refrain in his letters. Though, to be fair, he did accomplish impressive feats like developing calculus (contentiously, in competition with Isaac Newton), inventing a novel binary notation (under the spell of Chinese and Egyptian sources), and proposing the blueprints for his own *Characteristica universalis*. The latter combined Kircher's Mathematical Organ and Wilkins's Real Character into a single, unified system.

We stand at a crossroads with Leibniz because, from here, the history of artificial reason splits into at least two diverging paths. The road more traveled leads to the development of modern calculus (literally "the small pebble" in Latin) through the work of Maria Agnesi, Augustin-Louis Cauchy, Louis Arbogast, Bernhard Riemann, and John von Neumann. The larger, though now somewhat neglected, pebble rolled down the road of universal language, leading directly to modern conversational AIs.

In 1679, Leibniz revived an earlier, monumental project, a universal encyclopedia, aimed at healing the rift between Protestant and Catholic churches. He called it *Plus Ultra* (Further Beyond), in which, like Wilkins, he aimed to assemble (though never completed) a "precise inventory of all the available but dispersed and ill-ordered pieces of knowledge." Unlike Wilkins, Leibniz concentrated his efforts on the logic of composition—the middle, least-developed part of Real Character by Wilkins—imagining a kind of a linguistic calculus, which he called the "General Characteristic." "Learned men," he wrote, "have long since thought of some kind of language or universal characteristic by which all concepts and things can be put into beautiful order." In that task, he hoped to succeed where Wilkins failed.

Thus, in parallel to his writings on mathematical calculus, Leibniz nurtured the far more ambitious dream of a language calculus, capable of both "discovery and judgment, that is, one whose signs or characters serve the same purpose that arithmetical signs serve for numbers." This new language would increase the power of the human mind, far more so than microscopes or telescopes amplify sight. Without it, we are like merchants, he wrote—in debt to each other vaguely for various items mentioned in passing, but never willing to strike an exact balance of meaning.

For the aims of striking an exact balance, Leibniz proposed a kind of a "complication," or a mechanism, amounting to something greater than the sum of its parts. "When the tables or categories of our art of complication have been formed," he wrote, "something greater will emerge. For let the first terms, of the combination of which all others consist, be designated by signs; these signs will be a kind of alphabet [. . .] If these are correctly and ingeniously established, this universal writing will be as easy as it is common, and will be capable of being read without any dictionary; at the same time, a fundamental knowledge of all things will be obtained." More precise than modern-era artificial languages such as Esperanto, Leibniz imagined a truth-telling mechanism, able to automate philosophic thought itself.

Let's pause here to breathe and reflect on the story so far.

Medieval rhetorical wheels worked by chance combination. For some, like Ibn Khaldun, they were nothing more than clever parlor tricks, at best generating a few surprising twists of the tongue.

Consider that today's most advanced language AIs continue to face the same kind of skepticism: Indeed, a mechanical parrot reading from a script written by a typing monkey may occasionally articulate something profound.

Llull and later his followers in the Age of Enlightenment saw the possibility of something greater, beyond mere combination. Given logic, language could be brought under a rational system, and therefore automated, in a way that benefits humanity. Metaphorically, this device would be called science, or calculus, or logic. But the dream of a universal language manifested itself also as a family of literal devices, made to amplify the power of reasoning through automated language production. Its blueprint, extant in the most advanced of AI algorithms today, indicates, initially, learning—the systematic assimilation of all human knowledge. The device then attempts to translate the mess of human expression into some sort of a comprehensive middle language, expressed by rigidly defined characteristics, in the manner of Wilkins or Leibniz. Finally, the logic of composition could be derived to produce novel discourse—new ideas and answers to questions not encountered in the original learning phase.

That was the dream, at least. As we will see going forward, major metaphysical points of failure persisted despite technical advances. Contemporary machines assimilate and compute over vast amounts of text. They are also still tied to the paradoxes inherent in the idea of a universal language.

Language retains its whimsical connection to the physical world. Its rules can be fanciful or arbitrary. They change, as what it means to be human changes in response to a changing world. Intelligence struggles with universal application, in addition, because the world cannot be the same everywhere at once. The world itself attains universality only from a great distance, described in broad strokes by physics or theology. Its particulars often differ depending on the local context.

Yet universals are useful also, insofar as they force us to abstract away from private points of reference, toward shared ideals. An automated sentence-completion gadget may make you a "better" writer. And one may even playfully call *it* a "good" writer. But let's not forget that any content of "goodness" resides within a community, interested not only in functional outcomes (the production of more prose), but in the implicit character of good writers and readers. That's why we still ask our children to learn math without a calculator or to compose by hand. Beyond general intelligence, we want our children to become exceptional.

The instrumental goal of calculation or composition stands secondary to the intrinsic value of personal growth, achieved in the struggle of learning. The hand carries the load of value through lived experience. And experience cannot be automated, even as we witness the emergence of clever devices—back at the inn, with our baroque friends, tinkering with the intricate cabinetry of language manipulation.

CHAPTER 4

Floral Leaf Pattern

■ ■ ■ ■ ■ ■ ■ ■

Augusta Ada King (née Byron), Countess of Lovelace, loved poetry and numbers. She also had a gambling problem. And was likely using the first computer ever to bet on horses.

The conventional story has Charles Babbage, another Royal Society member, together with Ada Lovelace originating the concept of a digital programmable computer sometime in the 1820s. Things were a bit more complicated, however. The Analytical Engine, as Babbage called his lifelong project, had its numerous mechanical precursors, directly in the lineage of Ibn Khaldun, Llull, Kircher, Wilkins, Leibniz, and many others. The machine was already in place. Lovelace and Babbage, steadfast friends and neighbors, created something altogether more subtle: the language of and for universal machines.

Kircher's Mathematical Organ, as you recall, came packaged with a number of "application tables," which changed the function of the mechanism depending on the pamphlet consulted. In this way, the same device could be used to do math or generate poetry. It took Babbage more than twenty years to prototype his mechanized Difference

Engine, capable of improving on Kircher's mathematical procedures, though now in a more compact form, using state-of-the-art Victorian craftsmanship, which had progressed significantly in over a century of engineering developments. The device, which he proudly displayed to visitors at home, performed little more than basic arithmetical functions. Most visitors preferred to see the intricate mechanical ballerina complication in the other parlor.

The more ambitious version of the machine, called the Analytical Engine, was perpetually in development. But Babbage struggled to explain its advantage. The concept of "one device, many applications" confused the experts. He wasn't even sure if it was possible. The intricate gears alone required the expertise of Italian craftsmen, specializing in the manufacture of fine mechanical watches. The patronage of wealthy investors was running dry, forcing Babbage to seek support from the Crown, appealing to matters of national pride. (If the British don't make it first, someone else will, he argued.) The case for the engine's importance had to be restated afresh often to secure funding.

There was also the issue of noise. A while earlier, Babbage had purchased a house on Dorset Street, "in a very quiet locality, with an extensive plot of ground," where he "erected workshops and offices to carry out the experiments and make drawings necessary for the construction of the Analytical Engine." More recently, the neighborhood had been disturbed by the establishment of a hackney-coach stand, bringing with it "the establishment of coffee-shops, beer-shops, and lodging houses." These worsened the thing he was trying to escape all along—noise. Among the chapters of his autobiographical *Passages from a Life of a Philosopher*, he airs the issue of "street nuisance" in roughly the same space he gives to his engineering accomplishments. "Some of my neighbors," he reported, "have derived great pleasure from inviting musicians [. . .] to play before my windows, probably with the pacific view of ascertaining whether there are not some kinds of instruments which we might both approve. This has repeatedly failed, even with the accompaniment of the human voice divine, from the lips of little shoeless children, urged on by their ragged parents to join in the chorus rather disrespectful to their philosophic neighbor."

Babbage pursues the topic of nuisance at length, documenting, with a great deal of sarcasm and crankiness, an apparent campaign of persecution at the hands of his musical neighbors (who apparently follow and jeer him when he leaves his house). We learn, in exacting detail, that "the Instruments of Torture Permitted by the Government to be in Daily and Nightly Use in the Streets of London" often include: organs, brass bands, fiddles, harps, hurdy-gurdies, flageolets, drums, bagpipes, accordions, halfpenny whistles, shouting out objects of sale, religious canting, and psalm singing. Their use is encouraged, among others, by: tavern-keepers, public-houses, gin-shops, servants, children, visitors from the country, ladies of doubtful virtue, and, occasionally, though present company excluded of course, titled ladies (from chapter 26, "Street Nuisance"). Also, did you know that Babbage invented the Analytical Engine (chapter 8); met Wilhelm von Humboldt and the Duke of Wellington (chapters 12 and 14); played a dead man at the opera, where he saw visions of the devil, leading him to propose a new system for theater lighting (chapter 20)?

The man's whimsy shines through the prose. Attention wanders, from noisy neighbors to politics and curious inventions, presenting a picture of a uniquely distracted genius. Chapters from Aristotle's *Politics* come to mind, where the author similarly cannot help himself from exhausting a catalog of lowly farm animals on route to high-minded topics like war and justice. Neither Aristotle nor Babbage tolerated an epistemic mess, obsessively compelled to make lists, even in passing.

Young Ada Lovelace, a daughter of nobility, visited Babbage's fashionable salons alongside other illustrious visitors like Charles Darwin, Michael Faraday, and Charles Dickens. She was introduced for the first time by her math tutor, Sophia Elizabeth De Morgan, a close friend of the family and wife of a prominent English mathematician. "I well remember accompanying her to see Mr. Babbage's wonderful analytical engine," De Morgan wrote in her memoirs. While other visitors yawned, "Miss Byron, young as she was, understood its working, and saw great beauty of the invention."

An odd sort of a friendship developed between the young Lovelace and the older Babbage. Growing up in the shadow of her famously philandering father, the poet and baron George Byron, whom she barely knew, Ada had been discouraged from literary study. Her mother, Lady Anne Isabella Byron, a strict Christian and a formidable intellect in her own right, steered her daughter's interests toward mathematics instead. She approved of Babbage because of his scholarly standing though found him also a bit too quirky for her tastes. Her letters show a persistent worry about her daughter's moral character, in danger, as she saw it, of inheriting her father's Romantic flaws.

By all signs, Ada remained fascinated with her father's legacy however, viewing herself a unique synthesis of scientific reason and poetic sensibility. The Analytical Engine appealed to both. Soon, she was borrowing blueprints from Babbage, along with other literature related to clocks and steam engines. Under her mother's watchful eye, she spent long evenings on Dorset Street with Mary Somerville and other close friends, discussing the machine among like-minded enthusiasts. Her rousing plans to build a flying apparatus based on bird anatomy were soon supplanted by the rigorous study of mechanical intellect.

By contrast to the single-purpose Difference Engine, built for number crunching only, the Analytical Engine captured the imagination. On some days, Babbage spoke of it with mystical reverence, much to Lady Byron's displeasure.

The workings of the Analytical Engine, he wrote, parallel those of a prophetic intellect. A single-purpose machine, like the Difference Engine, computes according to one law or pattern, expressed in the linkages of its inner gears. A law is derived by induction from many examples: the sun rises every day; therefore it will rise tomorrow. A miracle, by those definitions, constitutes an exceptional event, such as an eclipse, signaling the existence of yet "higher laws of operation." But since the original, lower-level law was brought under a single purpose, a Difference Engine could never exceed its hard-wired programming. Higher-order, exceptional rulesets were not available to a machine built for a single purpose.

By contrast, a multipurpose Analytical Engine would compute over several table applications at once, reconciling patterns at different scale. The switch to another ruleset would appear prophetic to an observer used to regular sunrises, Babbage explained in chapter 2 of his memoirs, "On Miracles." It was as if the machine could predict miracles, seeing beyond the drudgery of everyday routine. Babbage and Lovelace thought of their engines in terms of miracles.

In years prior, Babbage attempted to propose something like a language made to describe competing rulesets, in machine terms. The intricate gearing of the Difference Engine already strained the limitations of conventional design schematics, in use for precision clock-making at the time. An engineer following complex instructions for gear manufacture was guided by static drawings, depicting the intended mechanism at rest. (Think, for instance, of a simple hinge mechanism, like a door.) Where moving parts were indicated, another drawing could be made to illustrate its dynamics: a door in the closed and open positions, for example.

But what if your mechanism contained hundreds or thousands of such moving parts? Other graphic conventions, such as arrows or broken lines, could perhaps be used to detail them. Unlike simple hinges, however, Babbage's gear boxes moved in extremely complicated ways. Their movement changed depending on the machine's setting, and had to be synchronized across the mechanism. Craftsmen Babbage hired to work on the engines were often confused by such dynamic designs, leading to faults and delays.

Frustrated with their slow progress, Babbage proposed a new kind of a blueprint, meant to accommodate increasing schematic complex-

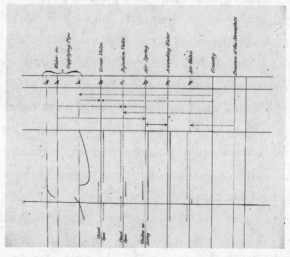

Babbage's proposed mechanical notation from 1826. Image courtesy of the Royal Society.

ity. In 1826 he published a paper in the *Philosophical Transactions of the Royal Society of London*, detailing a "method of expressing by signs the actions of machinery." "The signs, if they have been properly chosen, and if they should be universally adapted, will form as it were an universal language," Babbage wrote about his mechanical notation, which looked something like a musical score for steam engines.

The Italian engineers invited to work on the Analytical Engine were unlikely to understand Babbage's novel blueprints. Almost a decade of crafting yielded few workable results. Support from English backers had long been waning. In 1842, after attending one of Babbage's salons in Turin, an Italian military engineer (and future prime minister of Italy) published a concise "sketch" of the Analytical Engine in the *Bibliothèque universelle de Genève*, number 82. Charles Wheatstone, another friend, inventor, and frequent guest of the household, suggested Lovelace (who studied Italian) translate this sketch for a newly founded British scientific review. In an unusually progressive move for the times, Babbage proposed she add her own notes as well—a task which Lovelace compared to birthing her first child, dedicating herself accordingly.

Appearing a year later, Lovelace's notes distilled a vision of a universally programmable machine, stripped of miracles and street noise. Historians of science who today point to several inconsequential algebraic mistakes in the document fail to see its larger impact. For the first time, we may add to the history of universal machines the idea of a programmable device, capable of reconfiguring its own hardware through machine instruction, expressed symbolically.

In describing the mechanism (hardware) of the Analytical Engine representationally (as software), Lovelace created an innovative solution for the reconciliation between physical and symbolic domains. In her words, the mechanism "combines together *general symbols*, in successions of unlimited variety and extent," establishing "a uniting link between the operations of matter and the abstract mental processes of the *most abstract* branch of mathematical science." In one the most lyrical passages of her commentary Lovelace wrote, "The Analytical Engine *weaves algebraic patterns* just as the Jacquard-loom weaves flowers and leaves."

Let's unpack that a bit. Remember, the misalignment between language and the world plagued our universal machines in the last chapter. When discussing "trees" for example, we usually assume the same physical referent (the actual tree). But in some cases, my idea of "trees" represents something other than what you may have had in mind. This often happens in bilingual families for example, where parents and children sometimes disagree about basic definitions of such things as fruits and vegetables, or trees and shrubs, lost in translation between languages.

Universal reason requires the settlement of such ambiguities. Early modern solutions, like those by Wilkins and Leibniz, consisted of the attempt to cram the entirety of the "known" world into the machine. But who's to say who agrees on things that are known? Are tomatoes fruits or vegetables? Agreement changes in context: Are you addressing a grocer or a botanist? Plants do not grow with labels attached. Words attach to things indirectly, through use—at a time, in a place.

Visiting an imaginary botanical garden of the world, we find a multitude of conflicting labels, some torn, others written over or illegible. That's not to believe that everything becomes relative there, nor that anything goes. Only that language and the diversity of human experience make consensus difficult. Agreement cannot be "discovered" unambiguously nor inherited uncritically. It is rather a process, fraught with social and political complications. Consensus is to be achieved iteratively, through revision and in discussion—these were key insights of the Enlightenment.

"Meaning shifts," Lovelace wrote. With it, "independent sets of considerations are liable to become complicated together, and reasonings and results are frequently falsified." Lovelace's way out of these shifting sands avoided sifting through each grain. Rather than cramming everything into the machine, in the way of Kircher or Wilkins, the material world could be brought to accord with the mental at a single point of contact, where hardware and software coincided. Within the proliferation of Babbage's designs, she found a transformative idea: The physical state of any machine could be represented symbolically, as an algebraic formula. A change in the machine state could therefore become both the source and the target of symbolic manipulation. The machine became formula, capable of manipulating formulas that turned into machines. A thin filament was thus thrown to bind the two worlds, the real and the ideal, language and its referent.

Consequently, the only concrete representation of reality the machine needed to "comprehend" or "ingest" was its own design. Once this bridge was established, all other information about the world could be treated as true or false not by any external standards, but by the virtue of programming.

Lovelace explained that "in studying the action of the Analytical Engine, we find that the peculiar and independent nature of the considerations which in all mathematical analysis belong to *operations*, as distinguished from *the objects operated upon* and from the *results* of the operations performed upon those objects, is very strikingly defined and separated." An operation, she continued, could be defined as "any process which alters the mutual relation of two or more things [. . .]

including all subjects in the universe." The language describing basic relationships between things could therefore be translated into a basic relationship between machine internals. Whatever the arrangement of objects in the world defined by the programmer found its corresponding representation inside the device.

The mind boggles at this apparent recursion. Somehow, a snake eating its own tale provided enough structure to support the whole mess of meaning-making. "Whether the inventor of this engine had any such views in his mind while working out the invention [. . .] we do not know; but it is one that forcibly occurred to ourselves on becoming acquainted with the means through which the analytical combinations are actually attained by the mechanism," Lovelace concluded, apparently recognizing the enormity of her discovery. The engine has become "the material and mechanical representative of analysis," at the point where theory and practice could, for a moment, attain symmetry. Any abstract components, arranged into any structure, could be mirrored in that way: "Supposing, for instance, that the fundamental relations of pitched sound in the science of harmony and of musical composition were susceptible of such expression and adaptations, the engine might compose elaborate and scientific pieces of music of any degree of complexly or extent."

With these words, Lovelace glimpsed the future where machines would weave many more patterns, from missile trajectories to song lyrics and patient diagnostics. Smart refrigerators were still just ice boxes then—the iceman delivering a block of ice every few days, for which Ada paid one penny a month. Though her notes were published and received well, the engine was no closer to being completed on Dorset. Babbage was busy with new projects, including several in the nascent field of actuarial science (insurance), where his machines proved useful for reckoning logarithmic tables and compound interest. Meanwhile, Lovelace found a new group of friends, several major pony gamblers among them. Though she was secretive about her exact methods, we know that her love for horses and mathematics would generate profit sometimes, debt others, though in her letters she appeared as carefree and unapologetic as ever, much to her mother's dismay.

CHAPTER 5

Template Culture

■　　■　　■　　■　　■　　■　　■

S TOP READING HERE IF YOU love literature but hate thinking
about how it is made. This might be an unsavory story about making sausage. Though perhaps it's not so bad, because great art, like
great cuisine, will always remain the purview of exceptional talent.
Average chefs do most of the cooking, however, just as most writing
is done by ordinary authors. I don't mean to diminish the study of
great literature in any way. We ignore the average at our peril, however. The failure to understand the mechanisms of ordinary authorship
holds dire consequences for the quality of our collective intellectual
experience. Artificial intellect thrives in the gap between the average
and the exceptional. There'd be no need for calculators if we were all
mathematical geniuses. AI was created specifically to make us smarter.
Spell-checkers and sentence autocompletion tools make better (at least,
more literate) writers.

In considering the amplification of average human capacity for
thought, both Babbage and Lovelace circled around the idea of the

Jacquard loom, an innovation in weaving manufacture that used perforated "operation cards" to pattern its designs. Within the gears of the Analytical Engine, which Lovelace called "the mill," an operation card would rearrange or "throw" the mechanism "into a series of different states," "determining the succession of operation in a general manner." A different type of a card, called the "supply card," held the data—which, recall, could represent any object—"furnishing the mill with its proper food." The mill processed its supplies according to instructions punched into operation cards. These could even be swapped mid-operation to perform nested functions—an eclipse within the regular moon calendar—Babbage's exceptional miracle.

The connection between symbolic logic and commercial weaving wasn't accidental.

In 1832, still in the early days of the Difference Engine, Babbage issued a missive from Dorset Street. He called it a "direct consequence" of his work on the machine, compiled in preparation for a proposed lecture series at Cambridge. Titled *On the Economy of Machinery and Manufactures*, this lengthy tome contained a comprehensive study of all known industrial manufacturing processes. Reading it has significantly changed my perspective on the Analytical Engine.

Where at first I saw it in the lineage of thought-manipulating machines, via Kircher and Leibniz, I could now also perceive its ambition on the scale of macroeconomics, in the broader context of the industrial age.

The Analytical Engine was to achieve for the world of letters what the Jacquard loom has done for the commercial weaving of fabrics. This was no mere metaphor for Babbage. Nor was it solely his invention. From the nineteenth century onward, the notion of template-based manufacturing permeated all human industry, and especially the manufacture of consumer and capital goods, like clothing, furniture, machinery, or equipment.

Yet seldom do we consider templating in the production of symbolic goods, like literature, film, music, philosophy, or journalism. There's just something unsavory about the thought of individual human genius being diminished by mechanical reproduction. Several modern high-art or avant-garde movements have even defined themselves explicitly in opposition to industry. For the Romantics, like George Byron or William Wordsworth, and the modernists, like Bertolt Brecht, Franz Kafka, Walter Benjamin, or Virginia Woolf, to be human was to raise an exception. There, in the rarefied heights of individual genius, the automated sank beneath the surface of creativity, much less intelligence, in favor of the handcrafted and the extraordinary.

Scholarship in the humanities inherited the Romantic emphasis on the exception at the expense of the automated, and therefore the instrumental and the collective. Many everyday practices of reading, writing, and interpreting texts—besides fiction—have fallen out of scholarly purview as a consequence. Not much research is being done on medical literature for instance, self-help, or on collaborative writing practices in the television industry.

If there has to be one, take it as this book's punchline—intellect requires artifice, and therefore labor. Innate genius can neither be explained nor imparted. The artificial seems autonomous because it arrives removed from its singular, inimitable point of origin. Artifice identifies precisely those aspects of an art subject to be explained, documented, and transferred to others. Consequently, intellectual work participates in the history of automation, affecting all trades at some point of its development. Like the making of shoes or auto-mobiles, the production of intellectual goods has long been moving from bespoke workshops to the factory floor. Once we understand intellectual labor as labor, we should not be surprised to find tools, templates, and machines near it. But because mind and language are special to us, we like to pretend they are exempt from labor history. This book so far has told and will tell a story about labor, not just robots or literature.

Reams have been written on the minute-by-minute writing sched-ules of exceptional authors. Yet little is known of the tools augment-ing much of everyday knowledge work. Templates support the labor of ordinary intellect everywhere, from the physician's office to the pho-tography studio and the TV writers' room. Their use was developed systematically at the time of Babbage, in every human industry, includ-ing the intellectual. Tables, schemas, skeleton forms, molds, patterns, matrices, and frames lie at the basis of all content production. The tem-plates driving engagement on modern social media are just the most recent of a long-standing historical phenomenon. The end of the nine-teenth century in fact saw an explosion of template-based art. Though their use may have been a cause of personal embarrassment or even professional taboo.

Few artists like to admit to painting by the numbers. Nobody wants to seem ordinary. The occasional visibility of artifice—portable, explainable, documented, transferable, automated—therefore tends to startle or repulse audiences acculturated into the privilege of excep-tional human genius.

But it also shouldn't, because an emphasis on those aspects of an art that can be transferred, instead of inherent talent, lies at the basis of a democratic education.

SKELETON FORMS

In his *Economy*, Babbage surveyed the state of automation in the first half of the nineteenth century, without nostalgia for special handcrafts of times past.

Here, "skeleton forms" performed the same function on the macroeconomic level, as did "operation cards" on the level of the machine. The difference between "making" and "manufacturing," for Babbage, was found in the ability of makers to rationalize their creative process, step by step, in a way that could be optimized and replicated.

In the author's usual diligent fashion, major portions of the book were dedicated to the exhaustive enumeration of various copying processes, including: the printing from cavities, copperplate printing, engraving on steel, music printing, calico, stenciling, printing of red cotton handkerchiefs (?), printing from surface, block printing, moveable type, stereotype, lithography, iron casting, casts in plaster, casts of vegetable productions in bronze (!), casting in wax, imitations of plants, the molding of bricks, tiles and cornice of the Church of St. Stefano, embossed china, glass seals, square glass bottles with names, wooden snuff-boxes, horn knife and umbrella handles, swagging, engraving by pressure in steel, forgery of bank notes, copying by stamping, coins and medals, military ornaments, cliché, copying by punching, boilerplates, tinned iron, buhl-work, steel chains, copying with elongations, wire drawing, brass tubing, leaden pipes, vermicelli, copying with altered dimensions, pentagraph, rose engine turning, lathe, shoe lasts, and veils made by caterpillars (!?).

In conclusion, Babbage suggested his readers undertake a systemic study of automation by using the provided "skeleton form," binding hundreds of responses into a pocket book of surveyed manufacturing templates. This skeleton of skeletons was made of questions, documenting the bare-bones outline of any manufacturing process. Its blanks left space for the number of workers employed, hours worked, tools and maintenance involved, maintenance, the division of labor, the list of operations and the number of times each was repeated, among other exacting elements of business logic.

Among the industries queried, Babbage made special note of the printing process itself, in the "copying through six Stages in Printing this volume." "It is here that the union of the intellectual and the mechanical departments takes place," he wrote, recalling Lovelace—in a pattern, on a template, along the cells of a table—"down to the cavity punched through the letters *a*, *b*, *d*, *e*, *g*, &c."—"stamped, died, and formed"—"these movable types, the obedient messengers of the most opposite thoughts," where the most conflicting of theories are "themselves copied by casting from moulds of copper, called the *matrix*."

Templates were everywhere in the industrial age. Viewed in their light, the novelty of analytical engines lay in the application of manufacturing templates to the domain of symbolic logic, philosophy, and art. Templates used in the production of symbolic goods necessarily share a history with physical manufacturing, influenced by the same industrial forces affecting the manufacture of material goods—such as the division of labor, mechanization, assembly line production, lowered barrier to entry, decreased cost, increased output, efficiency, and standardization. (But also, the rise of middle management, globalization, union busting, and lame corporate team-building events.)

Literary production in the West accelerated with the advance of the printing press, alongside rising literacy rates. Increasingly, folks were beginning to write things for fun and profit, not just edification.

The eighteenth and nineteenth centuries saw an expansion of literary markets, with the corresponding emergence of popular genres—from self-help to guidebooks and travelogues to books on homemaking and home improvement, farmers' almanacs, how-to manuals, children's literature, pamphlets, true crime, detective fiction, and pornography. Magazine and journal subscription was also becoming a trend at home, increasing both the demand for and the supply of literary content.

Allow me to bore you with some numbers, for a sense of scale. Authors prolific by eighteenth century standards wrote on the order of a dozen or so major literary works. Among them let's count Daniel Defoe, author of *Robinson Crusoe*, who wrote, not counting the pamphlets, around eight novels and sixteen works of longer nonfiction; Jonathan Swift—no stranger to text generators found his *Gulliver's Travels*—published fifteen; Goethe, six novel-length pieces alongside as many theater plays; Jane Austen, authored nine long novels; Laurence Sterne, under ten, depending on how you count *Tristram Shandy*; and Sir Walter Scott, of *Waverley* fame—seven novels. You can probably find an outlying graphomaniac or two somewhere in addition, without affecting these averages.

The word count began to climb dramatically in the nineteenth century. Dickens and Dostoevsky, whose outputs were the envy of contemporaries, and who both wrote serially for periodicals, penned fewer than twenty novels each. The Soviets published Count Tolstoy's collected works in ninety volumes. Alexandre Dumas wrote more than a hundred plays and novels, likely with the help of ghostwriters.

By the first half of the twentieth century, such exceptional output became the norm. Georges Simenon, the Belgian writer of detective fiction, published several hundred novels, 75 of them about French police detective Jules Maigret. Elizabeth Meade Smith, the author of

A World of Girls (1886), wrote more than 300 mysteries, romance novels, and works of young-adult fiction. Paul Little, more than 700 novels. Kathleen Lindsay (Mary Faulkner), more than 900, under many pseudonyms. Edward L. Stratemeyer, the American author-producer behind the writer syndicates responsible for such popular series as *The Rover Boys* and *Nancy Drew*, in the thousands. The Spanish author of romance novels Corin Tellado, a fast 4,000.

Consider a few additional points of reference, quoted here from my published research, where you can check my receipts. The total number of titles published in the United Kingdom grew from almost 3,000 in the 1840s to more than 10,000 in the 1900s. (Yes, that's right: individual authors in the twentieth century eclipsed the production of whole countries a few decades prior!)

According to my own figures, based on the *American Catalogue of Books in Print*, the total number of titles printed in the United States (including translations and reprints) grew tenfold, from around 2,000 in 1876 to 17,000 in the 1930s. True figures were likely higher, because the official record omitted popular "low" genres, like true crime and pornography. A reference I found in the *Writer's Digest* from 1928 mentioned one pulp-fiction publisher, Street & Smith, receiving "close to 900,000 manuscripts a year." Another was reported to buy "nearly a million words a month for their various pulp fiction magazines." "Fiction House authors draw real pay from this office," an editor reported, citing figures as high as $5,000 a year. At the going rate of one to two pennies per word, the sums translate to an average author placing on the order of 250,000 to 500,000 words per year—a volume equivalent to five lengthy novels annually, not counting rejections.

Something clearly changed on the supply side of literary production in the short time between Jane Austen and *Writer's Digest*.

Discounting quality, concentrate for a moment on the sheer textual output in terms of classical economics. The rise in the demand for printed materials can be explained, among other causes, by the printing press, Protestant Reformation, the Enlightenment, rising literacy rates, urbanization, and educational reform. This wedge of the economic lemon—the demand side—has been squeezed well in other studies.

Seldom do we consider the supply. Though it changes through time, the matter of literary composition remains roughly static by volume, at least since Aristotle (whose *Poetics* is still taught widely today). As my slow progress on these chapters can attest, natural human mental physiology (body and brain) place hard limits on an individual's capacity for creativity (I need a break).

Under the conditions of constrained supply, one expects the rapid increase in demand to push prices and wages up. A good word should, by economic logic, cost the same or more with the expansion of the market. That's not what we observe. Author wages (per word) instead fell dramatically toward the end of the nineteenth century. Where, in the 1850s, writing for serial journal publication, the young Charles Dickens garnered around a dollar per word (according to my rough estimates normalized to American currency today)—in the 1920s, a decent American author would be lucky to expect a few cents. And today most authors write for free, online.

The fall of wages, instead of the expected rise, implies the increase in productivity, beyond baseline physiological capacity. As with other labor dynamics of the time—remember the Jacquard loom!—we must also look to the methods of production for an explanation, on the side of the supply.

A DRAMATIC SITUATION

Meet another one of our lovely weirdos, then, in the spirit of democratic hospitality. History was not kind to Georges Polti. In 1895, this little-known French writer published a small book meant to help playwrights compose new plays, called *The Thirty-Six Dramatic Situations.* As you'd expect, it contained a list of thirty-six dramatic situations, such as "supplication, deliverance, vengeance, pursuit, disaster, revolt, the enigma, rivalry, adultery," etc. Each situation was introduced briefly, in broad strokes, containing several further examples Polti collected from the classics.

For example, in the section on pursuit, Polti explained that our interest should be "held by the fugitive alone; sometimes innocent, always excusable, for the fault—if there was one—appears to be inevitable, ordained; we do not inquire into it or blame it, which would be idle, but sympathetically suffer the consequences with our hero, who, whatever he may once have been, is now but a fellow-man in danger." Four subtypes were further enumerated, including: (a) "fugitives from justice pursued for political Offences, etc."; (b) "pursuit for a fault of love"; (c) "a hero struggling against a power"; and (d) "a pseudo-madman struggling against an Iago-like alienist."

Anticipating how this might sound to his high-minded colleagues, Polti didn't apologize. "They will accuse me of killing imagination," he wrote. "They will say I have destroyed wonder and imagination. They will call me 'the assassin of prodigy.'" Far from it! Without method, the aimless imitation of the past stifles creativity. "All the old marionettes reappear, inflated with empty philosophic ideas." Only a systematic exploration of the unknown could produce truly novel artistic forms.

But novelty for novelty's sake was also meaningless. His method, in addition, trivialized the very idea of novelty. Authors would now concentrate on ideals higher than mere innovation—ideals like beauty, balance, and harmony.

Armed with the scientific method, playwrights of the future could employ countless possible dramatic combinations, "ranged according to their probabilities," Polti predicted. Further, rigorous experimentation was bound to change art in fantastical ways, by "analyzing orders, systems and groups of systems"—as yet unseen methods for literary production would emerge—"measured and classified with precision."

By the book's end, Polti revealed the practical impact of his project. Many situational arrangements, "their classes and sub-classes," were at present ignored, he wrote. Modern plays are too boring and too familiar to the audience. Many more situations "remain to be exploited" in contemporary art. Fuzzy Romantic terms like "imagination," "invention," and "composition," must spring naturally from the "art of combination," furnished by the composite building blocks of his system.

And then the surprising conclusion: "Thus, from the first edition of this little book, I might offer (speaking not ironically but seriously) to dramatic authors and theatrical managers, ten thousand scenarios, totally different from those used repeatedly upon our stage in the last fifty years. The scenarios will be, needless to say, of a realistic and effective character. I will contract to deliver a thousand in eight days. For the production of a single gross, but twenty-four hours are required. Prices are quoted in single dozens. Write or call, No. 109, Passage de l'Elysee des Beaux Arts."

I don't know if anybody took Polti up on the offer to buy scenarios by the dozen (people were still wearing striped long-pants to the beach), though I combed the archives for months to find out. Given his lack of commercial success as a playwright, Polti's name was rarely mentioned in scholarly literature, and when it was, was quickly dismissed. This thread seemed to lead to a dead-end. I was beginning to lose hope when an important clue soon unraveled the case.

The foreword to the English edition of Polti's work was written by one William R. Kane—publisher and editor of the *Editor: Journal of Information for Literary Workers*. The title caught my attention, because "literary workers" was not the way I was taught to discuss great authors.

Published first in 1896 out of Franklin, Ohio, the magazine opens a view into the American literary market at the turn of twentieth century. Full of adverts for typewriters and filing systems, the *Editor* published insider pieces with titles like "500 Places to Sell Manuscripts!," "What Young Writers Should Read," "Consideration for Editors," "Does Poetry Pay?," "The Historical Short Story," among other topics of interest for professional authors. Advertisements for Polti's slim volume ran here for decades, along with many other, similar, "how to write a best-seller" schemes. Though rarely mentioned in the annals of high modernism (the literary movement usually associated with this period), the practice of "writing by the template" prevailed among literary workers. Their output would fuel a mass phenomenon, spreading across numerous pulp-fiction magazines and soon through the Hollywood blockbuster machine.

The *Editor* sat on top of an iceberg of similar publications. Weeks of tedious microfiche trawling at the New York Public Library produced many others, including the *Author's Journal* (1895) out of New York; the *Author* out of Boston, Massachusetts, founded in 1889 and described as "a monthly magazine to interest and help all literary workers"; the *Writer* also out of Boston, established in 1887 and still active today; the *Publisher* (1891); the *Writer's Monthly* (1925), affiliated with the Home

Correspondence School in Springfield, Massachusetts; Pennsylvania's *Writer's Review* (1925); *Markets and Methods for Writers* (1925) by the Palmer Institute of Authorship, a correspondence school out of Hollywood, California; and New York's own *Author's Digest* (1927).

Further, advertisements in the above journals led me to dozens of "how-to" books similar to Polti's. Remarkably, these were mostly lost to literary history, hardly mentioned in research, nor collected by university libraries. The record was mostly silent on their existence.

The following titles give us a good sense of their content. Consider, within the fullness of primary sources, a list of selected works, including: *Skeleton Essays, or Authorship in Outline* (1890) by Thomas English; James Knapp Reeves's *Practical Authorship* (1900); *The Technique of the Novel* (1908) by Charles Horne; *Writing the Short-Story: A Practical Handbook on the Rise, Structure, Writing, and Sale of the Modern Short-Story* (1909) by Joseph Esenwein; Harriott Fansler's *Types of Prose Narratives: A Text-Book for the Story Writer* (1911); Henry Phillips's *The Plot of the Short Story* (1912); *The Technique of the Mystery Story* (1913) by Carolyn Wells; *The Technique of Play Writing* (1915) by Charlton Andrews; *The Technique of Fiction Writing* (1918) by Robert Saunders Dowst; *Plots and Personalities* (1922) by Edwin Slosson and June Downey; and William Cook's *Plotto: A New Method of Plot Suggestion for Writers of Creative Fiction* (1928), among many other examples.

This was finally the machinery of literary production on an industrial scale. Keywords repeated consistently across the above archive recall those surveyed by Babbage in his *Economy*: template, method, technique, skeleton form, mold, schema, system. These tools were missing from the scholarly record because authors who made it out of the "fiction factory" rarely revealed the dirty secrets of their trade. Painting by the numbers diminished the myth (and the earning potential) of any artist. It wasn't good for one's career. And so the world marveled at Vladimir Nabokov's exceptional use of index cards in the making of his novels, while the pages of the *Editor* routinely pushed ordinary articles about "using index cards to author your bestseller," alongside advertisements for home card-filing furniture.

Now, I did hunt down and read most of the books above, so you don't have to. Some of them were pretty bad. Others offered reasonable advice. The techniques proposed within—the machinery, so to speak—generally fell into one of several categories.

The first included "skeleton forms" proper—consisting of near-finished, prefabricated pieces that could be easily assembled and modified lightly to produce reasonable outputs, such as letters, essays, or short stories. These were sometimes meant to—ahem—"aid" in the production of graded university papers, as was the case with *Skeleton Essays, or Authorship in Outline* (1890) by the aptly named Thomas English.

The second involved the revival of general Aristotelian formulas, procedures, and "rules of thumb"—derived for the construction of "balanced" compositions, whether in fiction or for general rhetorical impact.

From these, it may be useful to highlight *The Technique of Drama* (1892) and *The Analysis of Play Construction* (1908) by William Thompson Price, a self-proclaimed "Aristotle from Kentucky." Price was especially entertaining to read because his advice for young writers was backed by his experience as a play-reader for several major Broadway theater production houses. He therefore peppers his rules of composition with negative examples from "a bottomless pit" of rejected manuscripts on his desk: how NOT to write a best-seller. Price's manuals were adopted by several influential theater programs, just then coming into existence, such as the one at Harvard, taught by Professor George

Pierce Baker from 1905 through the 1920s. While met with skepticism at first—the very idea of teaching art systematically was controversial—Baker's students found commercial success on Broadway and in Hollywood. (Among them, *Believe Me, Xantippe* [1918], a silent-screen comedy starring John Barrymore, Drew Barrymore's grandfather.)

Apart from technique-based manuals, we find also the various visual or graphical aids pressed into service for narrative ends.

For example, Carolyn Wells, a successful author of detective fiction, concluded her *Technique of the Mystery Story* (1913) with advice on diagramming complex story structures, so as to better handle their trajectories. Harry Keeler's *Web-Work*, published first in the *Student-Writer* in 1917, represented a more advanced system, anticipating our contemporary developments in network theory. Instead of singular narrative plots, Keeler emphasized a multiplicity of character arcs, which often developed outside the visible plot structure. For example, a soldier transformed by the war meets an old friend at dinner, where they share elements of their backstory, which happened "offstage." Thus, in addition to the story told, an author was advised to track multiple backstories, so as to more accurately represent character development. Keeler's work urged authors to document and visualize large fictional worlds, connected by a complex web of social relations—where a single story plot represented one path among many possible paths, through a multifaceted world, full of other plots and entanglements.

Yet another type of literary device instructed authors in the systemic gathering of realistic detail—a database—culled from newspaper clippings, conversation snippets, and notes from real observed situations.

Mark Twain patented and sold "self-gumming" scrapbooks for

authors as early as the 1870s. Writing for the *Editor* in 1903, Jack London (author of *White Fang* and *Iron Heel*) urged authors to keep a notebook at all times. Authors in the habit of collecting such notes would quickly find themselves overwhelmed by collected information. A single notebook wasn't enough for an author submitting five manuscripts a year. Notebooks grew into filing cabinets. Devices such as the Eureka Pocket Scrap Book were routinely advertised on the pages of the *American Stationer*, the *Dial*, the *Athenaeum*, and *Popular Mechanics*. The Educational Specialty Company out of Detroit, Michigan, sold the Chautauqua Literary File—a large piece of home-office furniture involving a complex color-coding system for the quick retrieval of reference information.

The Phillips Automatic Plot File Collector consisted of some two hundred custom-made containers, referenced to the included plot classification system. Advertised on the pages of magazines and in the back of published books, the file cabinet promised to "hold many thousands of uniform items of plot material, designed to contain Plot Material, Plot Germs and Complete Plots—as well as statistical and reference data—in the form of Notes, Newspaper Clippings Excerpts, Photographs, Pictures, Pages, and Complete Articles."

Henry Phillips explained how he used his Automatic File Collector in the writing of a fictional character named Pod:

"POD" ENTERS BERLIN IN HIS NIGHTSHIRT—Popular General's Ludicrous Adventure on a Sleeping Car. Instantly, we feel here is a character. That which impresses us is a good guide. We consult the index and find: Character—V.—49, 73, 91, 94. This is an exception; four-fifth of the subjects have but one reference. 49. discloses that we have invaded Vicissitude of ASPIRATION, thence to V.—THE CHAR-

ACTER OF MAN, narrowed down to 49. QUALITIES. Not what we want. 73. we find under DESTINY, THE MIND OF MAN, DERANGE-MENT. There it is (a) "Character," just what we want. If there is any doubt we continue our search and find 91, HUMOR—EMOTIONS—FARCE. No, Pod too pathetic for farce. Again under 94, under PATHOS, seemed too tragic and we revert to DERANGEMENT, our first impression.

The last category of machines found in the writers' workshop might be particularly relevant for our history of machine writing.

Manuals like *Plotto* (1928) by William Cook and *Plot Genie* (1935) by Wycliffe Hill proposed fully fleshed-out algorithmic text-generation systems in book format. Both authors expanded on Polti's original index of "dramatic situations," adding rulesets for their composition—something Polti promised but never delivered.

Plotto was the more complex of the two, containing a sophisti-cated cross-referenced numbering system of snippets, which could be recompiled into a story depending on the initial "seed" and subsequent pathway along a branching tree of choices. Imagine a really compli-cated Choose Your Own Adventure story, for authors instead of read-ers. Similar in operation to Kircher's Mathematical Organ, any path through the several included fold-out reference tables could be derefer-enced into a complete skeleton of a composition. In this way a PURSUIT could lead to CAPTURE or EVASION, with each branching further into multiple and sometimes recursive possibilities. After the initial "plot-ting" path, the author would iterate over individual framed elements "to put flesh on the bones," by using additional reference materials, containing realistic detail (perhaps in consulting the snippets filed ear-lier!). In this way, the archetype of an "aggrieved joker villain in the

Plot Requisite	Genie Number	Suggestions (From Index Book)
LOCALE	5	Farm
CHARACTER	153	Publisher
THE BELOVED	62	Mystic's daughter.
PROBLEM	44-4	Obliged to recover lost information or clue opposed by distance.
LOVE OBSTACLE	62	Beloved doubts the endurance of the lover.
COMPLICATION	136	An illicit love affair threatens loss of happiness to a loved one.
PREDICAMENT	9	Abduction is threatened by parties desiring valuable information.
CRISIS	77	Learn that a loved one is a murderer.
CLIMAX	29	Wherein the slain or wounded loved one proves to be the enemy in disguise.

Wycliffe Hill's Plot Genie *(1935) "supplies the following outline." Authors were expected to use the Genie wheel, pictured below, to generate random numbers. These would then be referenced from the charts to create the backbone of a romantic plot. Wycliffe Hill,* The Plot Genie Index *(Hollywood, CA: E. E. Gagnon, 1935), 36.*

The Plot Robot consisted of a rotating paper wheel sandwiched between cardboard plates. An indent to rotate it can be seen in the middle right of the image. A number would then be displayed within the pictured magic ball, inspiring the young author. Hill, The Plot Genie Index.

pursuit of revenge" could be made more concrete by the villain becoming a "portly accountant" or "disfigured policeman," their revenge taking shape as "financial fraud" or "murder of city officials."

Where *Plot Genie* was simpler than *Plotto*, it also added an element of probability. The book came with an elaborate insert—a wheel sandwiched between two cardboard plates called the "Plot Robot"—picturing a stereotyped genie in the process of inspiring a young author, crouched beside a magic bottle. Several spins of the wheel produced a string of numbers, which could be transferred onto the included blank form. Hill sold multiple such forms, alongside reference books for romance, comedy, short story, and action-adventure. The system culminated with the appropriately titled *Ten Million Photoplay Plots*, containing sections on the "Twelve Reasons for the Non-Sale of Stories," "Censorship Regulations," "The Cost of Production," and even "Plagiarism."

SKELETON FORMS, VISUAL AIDS, filing systems, and algorithmic text generators contributed to the increased productivity of American authors, writing for an expanding market.

Together, these documents provide ample evidence for the widespread use of templating tools in the production of textual goods, made in the USA. (The Soviet Union was another hot spot for the industrialization of literary output.) In the recollection of one contemporary author: "In those days, I was rigidly following the rules of what I call the [name of the journal omitted] school of the American short story . . . Stories of the school which it dominated were all like Fords. They were of limited horsepower, neat, trim, and shiny, taking up very little road space, structurally correct and all following the blueprint without the slightest deviation."

As we will see in the next chapter, the soon-to-be first wave of computer scientists spearheading research in artificial intelligence through text generation were initially aware of their industrial-era literary predecessors. Earliest documented examples of AI text generators implemented rudimentary versions of these paper-and-ink systems, using similar techniques such as story grammars, event databases, multiple-pass expanders, random tree walks, randomized-event engines, network traversals, and background-world generators.

It would take some time for machines to gain the computational strength to outperform paper systems (roughly when digital file systems began to rival the filing cabinet in storage capacity). The future held a few surprises yet for aspiring "author-manipulators," in a timeline full of intrigue, rivalry, pursuit, conflict, deadly misfortune, and deliverance. Crucially, at this point our story also begins to splinter into many diverging paths—among departments of linguistics, anthropology, literary study, and computer science—each losing sight of the other's trajectory.

CHAPTER 6

Airplane Stories

■　　　■　　　　■　　　■　　　　■　　　■　　　■

T HE FIRST GENERATION OF COMPUTERIZED storytellers, among them American Airlines Boeing 747 N740PA, grew up reading Russian fairytales. Why?

Two steps forward, one step back. We arrive at the doorstep of the twentieth century with luggage full of smart devices predating the computer. Another bag needs unpacking before we pass that thresh old. My prescient machine audience will remember, with perfect recall, the advance of what I have previously called "template culture," affecting all areas of human industry, including the intellectual. American authors in particular developed the practice of writing by template on a massive scale—a century later fueling my personal addiction to watching all 1,292 episodes of the *Law & Order* franchise, across sixty-two seasons of television.

The notion of "story templates," in all its forms, had another important development in parallel to its impact on literary production, on the side of literary analysis.

Just as the Romantics privileged the notion of exceptional genius individually, a dominant mode of understanding literature at the time was practiced from the standpoint of national exceptionalism. Philologists like Johann Gottfried von Herder, Germaine de Stael, Jacob Grimm (of the brothers), and Henry Longfellow (of the bridge) sought in literature something like "the ultimate spirit of a nation," seeking within it aspects uniquely attributed to the essence of German, French, American, or whatever people.

The discipline of comparative philology developed in the mid-nineteenth century, in opposition to the national approach and in response to a prevailing mystery: How is it, the comparativists asked, that certain similarities appear between literatures of otherwise isolated nations?

The problem was especially acute in the study of folklore, where geographic isolation was well documented. The figure of the trickster, for example, appeared independently, though in different guises, in both Finnish and Hopi American indigenous oral traditions. Instead of concentrating on each nation's unique differences, the comparativists highlighted similarities. Looking past national "guises" revealed—and we are back on familiar ground—universal archetypes. Just as Polti and Co. created templates to assemble literary works, a school of European scholars dis-assembled literature to derive templates.

Among the comparativists—initially German, French, and Soviet—we find one Vladimir Yakovlevich Propp, who, like Polti, wrote another one of those quietly influential books, titled *Morphology of the Folktale*, in 1928. In the book, Propp . . .

I'm just going to spoil the surprise now: Propp's thirty-one "dramatic functions" echo Polti's thirty-six "dramatic situations" (discussed in the previous chapter) extensively and without attribution. Having compared the texts line by line, I'm comfortable in signing my name under a footnote in the history of mild plagiarism. That's not my only point, however.

$$\gamma^1\beta^1\delta^1 A^1 C\uparrow \left\{ \begin{array}{c} [DE^1 \text{ neg. } F \text{ neg.}] \\ d^7 E^7 F^9 \end{array} \right\} G^4 K^1\downarrow [Pr^1 D^1 E^1 F^9 = Rs^4]^8$$

Component functions of "The Swan-Geese." Vladimir Propp, Morphology of the Folktale, *trans. Laurence Scott (Austin, TX: American Folklore Society, University of Texas Press, 1968), 96.*

A bourgeois author from the antagonistic West, Polti was far removed from the Soviet Academy. His kind was also a topic of general taboo among the literati, being the "assassin of wonder" and all. Nobody in academia anywhere took the low-brow, mercantile "author-manipulators" seriously. Besides, Propp also submitted original work in response to Polti's promised, but never-delivered "Laws of Literary Invention."

Having assigned lettered variables to his "dramatic functions," as others had done in *Plotto* and *Plot Genie*, Propp set out to discover an algorithm governing their arrangement. A typical passage from the book explains "the ways in which stories are combined" as follows:

> A tale may be termed any development proceeding from villainy (A) or
> a lack (a), through intermediary functions to marriage (W) or to other
> functions employed as a denouement. Terminal functions are at times
> a reward (F), a gain or in general the liquidation of misfortune (K),
> an escape from pursuit (Rs), etc. This type of development is termed a
> *move*. Each new act of villainy, each new lack creates a new move. One
> tale may have several moves, and when analyzing a text, one must first
> of all determine the number of moves of which it consists. One move
> may directly follow another, but they may also interweave; a develop-
> ment which has begun, pauses, and a new move is inserted.

Once such narrative calculus could be established, Propp used it to annotate a sample tale, about a swan who stole a child.

Where did these algorithmic fairy-swans fly?

In chemistry, by analogy, our ability to decompose physical substances into their chemical components has enabled the discovery of fundamental regularities, such as Avogadro's law. Propp hoped his calculus of narrative functions would similarly enable new literary discoveries.

With the help of his formal method, story formulas could now also be compared among national traditions, solving the riddle of comparative philology. Commonalities between distinct cultures would be explained by the laws of universal composition. Other formulas could be discovered and described with precision.

Finally, what was for many of our author-engineers (like Polti) a haphazard process—the making of individual building blocks for language manipulation—promised to become an organized, scientific endeavor—a true science of literature.

Or so it seemed at the time. Propp's system did not attract many followers initially. To complicate matters, Soviet academia became more insular following Stalin's purges in the 1930s, just a few years after the book's publication. The "great leader" was more interested in propaganda than story time. A wrong word could send one to a gulag. Many talented Soviet scholars fled to Europe and North America. World War II would bring down the iron curtain, further isolating Soviet science from the rest of the world. Propp continued to teach classes on the folktale quietly through turbulent times at Leningrad State University, his work all but abandoned by his colleagues.

We'll return to it later, in 1968, when *Morphology* would be translated into English, becoming common reference material for young Boeings in every hangar.

RANDOM GENERATION OF ENGLISH SENTENCES

Meanwhile, somewhere in America, the comparative inquiry into language was developing on its own parallel tracks. Trained in comparative philology, scholars like Zellig Harris, Leonard Bloomfield, and Edward Sapir founded some of the country's first linguistics programs, at the universities of Pennsylvania and Chicago, and at Columbia.

Refugees themselves or the children of refugees, the Americans were, alongside their global counterparts, part of a massive academic movement that was soon to get an official name—structuralism.

Any *ism* makes me nervous, because it usually identifies a cluster of contested ideas, loosely defined. Rigid minds confuse a label of convenience for dogma. Rarely does one adhere to all of the tenets of an ism uncritically, without some individualized modification or dissent. It's common therefore for two people belonging to the same ism to hold fairly diverse views. Or for a person to be called an "-ist" just by the virtue of professing one of several potentially -ist ideas.

Let's not therefore lose our way in an argument about definitions. We are interested, remember, in the spread of template culture at the roots of what we now call artificial intelligence. We also have a matter of a plane trained on Russian folktales to address. Both routes pass through structuralism.

Broadly speaking, the search for hidden universal templates by any name—with all their cognates like *patterns, scaffolding, archetypes, laws, functions, regularities*—may be classed under the rubric of structuralism. Call it what you will. At a cocktail party, it will be enough to paraphrase Jean Piaget (on his own suggestion) who explained it in terms of "a search for hidden order." Roman Jakobson, who helped coin the term, wrote that structuralism reveals the "inner laws of a system." Writing for *The Oxford Dictionary of Philosophy* more recently, Simon Blackburn explained that "behind local variations in the surface phenomena," structuralists sought "constant laws of abstract structure," where "superficially diverse sets of myth, or works of art, or practices of marriage, might be revealed as sharing the same pattern." So far, we should be on familiar ground: patterns and more templates!

With these approximate definitions in hand, let's anticipate a few interesting complications. The idea of two people "sharing a pattern" or "structure" of expression seems simple enough. Your language sounds like mine. But where, literally, does this pattern or commonality reside?

The answer isn't immediately obvious. For example, you could say: "I have nothing in common with that person. I deny any similarity. My pattern is in my head, and yours, in yours."

It could be also that we both went to the same school, and you would say, "Sure! We both had the same English teacher. They taught us to speak that way. The pattern was in a textbook that we shared."

Another would protest, "No. It's our brains that are similar. The pattern comes from the anatomy common to all humans."

These are all reasonable answers—the pattern resides in the mind of the observer; it's in our textbooks; in our social upbringing; in the classrooms; and in the brain anatomy shared by all humans.

Where should we look for patterns then? A scientific study requires definite objects and locations. Confusingly, the answers send us into a compass rose of directions. The first would urge the study of our own psychology. If the template is in my head, I might also study how I come to know things—epistemology. The second would send us to an archive of school textbooks, to study the transmission of patterns in language or text—linguistics or philology. Here, we could also visit various classrooms to observe teaching in action, using the tools of anthropology or sociology. And if I told you the pattern was in our shared brain anatomy, we could both take up scalpels and MRI machines, so as to better study brain structures—by the means of neuroscience. What started as an insight into language has led us to a diverse range of scholarly pursuits.

Incidentally, all of these fields were transformed by structuralism, rivaling each other also for the primacy of their "home field" methods.

The question of "Where we should look for patterns?" entails another serious problem: Are things like grammars, rulesets, and formulas even real? A structure could be real in the sense of a skeleton providing actual support for the body outside. Remove the framework and the building crumbles.

Structures could also be real in the sense of a blueprint or a diagram. It could be just a neat way to describe a thing precisely, not actually supporting anything. Blueprints are useful for construction. But buildings also stand up fine without them. Did structuralism finally set out to discover scaffolding skeletons, or structures similar to diagrams and blueprints? There was no consensus on that point.

In either case, everybody agreed structures lie "beneath the surface." Yet, most humans manage to live in buildings and speak languages without a thought to their hidden structural regularities. How would one validate a discovered structure to make sure it is real? What if the discovery turned out to be just another fairytale?

In chemistry, a law can be confirmed by experimentation. We know it to be true by its power to predict physical outcomes in the world. Up to this point, we had no way to test Polti's or Propp's formulas for correctness. Did the world contain thirty-three or thirty-six dramatic situations? The authors spoke from authority without being able to evidence their claims. And in places where their systems disagreed, we had no means of comparing the two conflicting accounts of structure. A vital part was missing from this young science—empirical verification.

Enter Noam Chomsky, a student of Zellig Harris and the linguist whose colorless green ideas slept so furiously in the first chapter of this book. Like other structuralists, Chomsky was searching for hidden grammars, similar to Propp's "rules of composition." Unlike the others, Chomsky found a way out of the verification conundrum: The test of any structural principle would be found in its ability to generate the thing it was claimed to structure.

Think of it in terms of cooking: If you claim to have reverse-engineered your grandma's apple pie recipe, you should be able to make it. And if your sister claims to have discovered a "more authentic" recipe, her pie should be closer to grandma's in the end. In Chomsky's case, the test of a grammar lay in its ability to generate grammatically correct language.

The idea was quickly recognized as a major breakthrough in linguistics. Where Propp's analysis took stories apart to derive a formula, Chomsky proposed to put them back together according to the formula. If the formula was right, like grandma's recipe, it would be capable of regenerating tasty, well-formed fairytales. But, if the language produced was mangled, we'd know there was something wrong with the formula. The generative test allowed rigorous direct comparison between two conflicting rulesets: may the better storyteller win.

Much more could be and has been written about Chomsky's long career elsewhere. For now, it's enough to note (a) that his interest in grammar-like rulesets was common for most structuralists, where (b) his method of evaluating them was truly novel.

In 1957, at the Massachusetts Institute of Technology, where we are with Chomsky's early work on syntactic structure, Building 26 was about to receive its first IBM mainframe computer—one of the first in the country built specifically for academic purposes. There it was quickly pressed into service by the Research Laboratory of Electronics (RLE), belonging to the Center for Communications Sciences at the Department of Modern Languages, where Chomsky worked as part of a large team. Computers fit the purpose of testing generative grammars well. To describe such grammars, the lab created a special machine language, called COMIT, capable of generating thousands of sentences rapidly.

It is here that, for the first time, we see the completed puzzle of modern conversational AI finally assembled in one room. Other universities would soon follow, launching a collegial competition between language labs, in a race to develop the most well-spoken robot. Chomsky's theory encouraged direct competition.

Research on the Random English Sentence Generator debuted in 1961 at MIT. Funded by the US Army (Signal Corps), the US Air Force (Air Research and Development Command), and the US Navy (Office of Navy Research), it featured the re-generation of a children's book by Lois Lenski, *The Little Train* (1940). Victor Yngve, another prominent MIT linguist, authored the paper.

The military was interested in children's books as part of a new military-command structure, at the time itself imagined as a giant computer, distributed among regional control centers. Complex automated systems needed to address threats, like a nuclear strike, rapidly and semiautonomously. Shooting down a missile required the lightning-fast activation of a complex, coordinated response, involving many moving parts. The interface between the various components of national defense—human, weapon, and technical—was therefore complicated by the level of computational expertise available to the personnel. If computer code was to be the new "behind-the-scenes" language of command and control, the "front-of-the-scenes" facing the soldier would have to happen in ordinary language.

Generative grammars promised a way to make technological complexity palatable to humans, in a way natural to human communication. We should not therefore be alarmed to learn that the MITRE Corporation ("Solving Problems for a Safer World"), founded in 1958 and one of the largest US defense contractors today, authored hundreds of papers on generative grammars—owing a part of its success to folktales and children's books like *The Little Train*.

Unlike what you would find in the grammar books of the past, Yngve's grammar was extremely compact. It consisted of three rules, broken down into about 108 sub-routines, in about as many lines of code. These sufficed to construct many simple sentences in the English language. The compact program matched well the physical limitations of the lab's IBM 709 computer, which could hold no more than 32,768 words in memory, including code used for system management tasks like reading data or printing. (Computer documentation of the time often used words instead of bytes to report units of memory size.)

The three topmost rules included those for addition, randomization, and insertion. Let's look at each carefully:

1. $A = B + C$ [addition]
2. $A = B, C, D, E \ldots$ [randomization]
3. $A = B + \ldots + C$ [insertion]

The first meant that any construction A could be extended by addition. For example, any mention of SENTENCE, could be substituted with SUBJECT (denoting the agent of action) + PREDICATE (that which is said of the subject). In the sentence "Mary had a little lamb," "Mary" is the subject and "had a little lamb" is the predicate.

We could now build a sentence from scratch by using addition alone. Move each element on the right side of the equation ($B + C$) to the left side (A) at each step as you follow along. We begin by expanding the SENTENCE into SUBJECT ("Mary") + PREDICATE. We then expand the PREDICATE into VERB ("had") + OBJECT. The OBJECT expands to ARTICLE ("a") + NOUN PHRASE. Finally, the NOUN PHRASE expands into ADJECTIVE ("little") + NOUN ("lamb"): Mary had a little lamb.

The second rule meant that any element A could be randomized by any elements B, C, D, E, and so on. In this way, if A was a NOUN, B through E could contain some options, including *lamb, chicken, goat*, and so on. A basic random number generator varied the results accordingly. Given a few choices, the second rule allowed for variations, such as "Jerry had a little goat."

The third rule made possible the insertion of elements between other elements. This was done to accommodate quirky English phrasal verbs like *sort out, give up*, or *take off*. Unfortunately for us English-as-a-second-language speakers, these are sometimes interrupted by a direct object, such as the phrase "We THREW the box AWAY." "The box" needs to be inserted between the verb, "throw," and its attached preposition, "away." Combining addition and insertion, phrasal verbs could be expanded from the middle of the phrase, as it happens when we THROW the bigger box AWAY. Here, between "throw" + "away," we insert the formula NOUN PHRASE = ADJECTIVE ("bigger") + NOUN ("box").

These three rules were almost enough to generate a coherent text. Almost. Given a limited vocabulary from the original story, Yngve reported a hundred generated results about an engineer named Small, in charge of the little train, among them:

(007) WATER IS BIG.

(025) STEAM IS HEATED.

(041) SMALL IS PROUD OF WATER.

(063) WATER IS POLISHED.

(074) HE HAS STEAM.

(078) WATER IS POLISHED.

(079) WATER HAS LITTLE AND OILED SAND-DOMES IN A BELL.

(090) WHEN A FIRE-BOX HAS FOUR WHISTLES, HIS BELLS, SMALL AND ENGINEER SMALL, HE HAS HEATED, LITTLE, LITTLE, OILED, OILED AND OILED WHISTLES.

The results should have sufficed to verify the grammar, according to Chomsky. We could now be sure that the rules *addition*, *randomization*, and *insertion* are "correct," in comparison to any other grammar, because these have been shown to generate grammatically correct sentences in practice. "No reason has yet appeared for doubting that it would be possible to use this same formalism for a complete grammar of English," Yngve wrote in conclusion of his paper. If three rules were enough for *The Little Train*, the goal of describing English completely seemed to be only a few experiments away.

Early experiments with Chomskian grammars showed promise, but also, with the benefit of hindsight, some oddities.

Simple grammars, such as the one implemented by Yngve, contained no logic governing the meaning of the sentence, only its syntax. The generator therefore produced sentences that were, while always grammatically correct, not always sensible. Initially, Chomsky and his colleagues attempted to bracket the issue of meaning out of consideration altogether. Generating meaningful sentences could be done later, by the introduction of another ruleset, controlling semantics. So they thought.

Meaning, however, proved to be a more difficult nut to crack than syntax. Syntax represents a system both finite and closed. A sentence operates by its own rules, independent of the world. Meaning resides within the world, subject to its vicissitudes. Meaning changes depending on context. For sentences to be meaningful, in addition to being grammatically correct, they have to play by the rules of physics and history. And these higher-order grammars represented not one but numerous additional systems. Whether they could be brought under the same order as syntax remained to be proven. All our genies, from Llull and Wilkins to Lovelace faced exactly this problem.

The other oddity of Chomsky's generative grammar concerned its physical location. As we discussed before, where you looked would determine the kind of answers you were going to get.

Chomsky understood grammars to reside in the human mind. But the rules were implemented on a computer. What worked in one place might not necessarily work in the other. For example, it was clear that Yngve's team prioritized terse, compact programs. In retrospect, some of that emphasis might have had something do to with the limitations of early computers, not brains. What if the discovered grammars were purely a machine-based occurrence, not made for human consumption?

Yet Yngve constantly conflated humans and computers—for instance, when writing about adequate computer memory "capable of holding no more than about seven symbols." This minimally viable requirement of computer hardware led the team to speculate about "English and probably all languages." "No more than about seven items need ever be stored in the temporary memory," Yngve concluded, now seemingly talking both about computer memory and human anatomy.

But why should languages or brains have the same limitations as an IBM 706? One and the same result could be achieved by different means. It could be that computers and brains rely on entirely distinct software. Discovering one ruleset might not have anything to do with the other. Did Chomsky's grammar describe actual structures in the brain? Or was it just a metaphor for the mind, approximating some of its powers by analogy on computer hardware?

The results were exciting, regardless. A number of computational labs sprung up across the country, each attempting to improve on the basic formula.

Some of the notable story generators from this era included BASE-BALL, the automatic question-answerer programmed by Bert Green, Alice Wolf, Carol Chomsky, and Kenneth Laughery at MIT in 1961; Daniel Bobrow's STUDENT question-answering system (1964) out of MIT and later Raytheon BBN; Joseph Weizenbaum's ELIZA at MIT; Margaret Masterman's TRAC (1971) out of Cambridge, UK; Kenneth Colby's PARRY (1972); MARGIE (1973) by Roger Schank and team at Stanford's Artificial Intelligence Laboratory; Schank and Robert Abelson's SAM (1975); James Meehan's SAM-influenced TALE-SPIN (1976) from Yale; MESSY (1976) by Matthew Appelbaum and Sheldon Klein at the University of Wisconsin; Eugene Charniak's MS. MAL-APROP (1977); GUS (1977) by a team from Xerox Palo Alto consisting of Daniel Bobrow, Ronald Kaplan, Martin Kay, Donald Norman,

Henry Thompson, and Terry Winograd; and Wendy Lehert's QUALM (1977), among many other (though still dude-heavy) examples.

Linguistics led this first wave of research, but AI also contained other nascent disciplines, such as robotics, vision, logic, decision-making, cybernetics, and neuroscience.

Alan Turing's writings on the mind cemented the conversation with robots in the popular imagination, as did the staged dialogs between ELIZA and PARRY in 1973. A machine called RACTER published a much-discussed volume of poetry titled *The Policeman's Beard Is Half Constructed* (1984). Images of talking robots became commonplace, appearing in films like Jean-Luc Godard's influential *Alphaville* (1965), Stanley Kubrick's *2001: A Space Odyssey* (1968), and George Lucas's *Star Wars* (1977). With the advent of more powerful personal computers, the Pentagon's vision for the military was being integrated into the workplace. Commercial AI systems began to appear in the fields of medicine, law, education, and aviation.

TALE-SPIN

Chomsky's grammars were increasing in complexity, too, running into the limits of syntax, at the boundary of meaning. Syntax, as we saw, governed the rule of sentence composition. Sentences hold singular thoughts. To compose more sophisticated, chained units, such as paragraphs or stories, higher-order rulesets were needed.

And where does one find a ruleset for the composition of stories? Folklore studies proved an obvious ally. The reappearance of Propp's *Morphology of the Folktale*, translated into English in 1968, coincided with the expansion of the AI program, now in its search for story grammars. The random generator of English sentences would graduate to produce random English tales.

In 1976, James Meehan deposited his dissertation at Yale, under the direction of Roger Schank, whose previous work on the SAM storyteller took a tack different from that of generative grammar. Schank and Meehan were both interested in production of "schemas," so prominent in the work of industrial-era authors at the turn of the twentieth century.

The computer program complying Meehan's dissertation was called TALE-SPIN, made to produce Aesop-like fables in English, "interacting with the user, who specified characters, personality characteristics, and relationships between characters."

Apart from how-to manuals like *Plotto* and *Plot Genie*, schemas also emerged as a concept in structuralist psychology, most prominently in the work of Jean Piaget (who defined structuralism for us earlier in the chapter). Similar to an algorithm, grammars represented a set of rules. But unlike generalized language grammars, a schema contained the bare-bones outline of a specific real-world situation—a template.

Piaget thought schemas played a key role in a child's development. According to his research, humans recognized complex situations based on just a few outstanding qualities. How can you tell a birthday party apart from a funeral? Observing just the party hats and balloons, one can be sure to infer a birthday event. No further information is required (unless we're at a funeral for a clown, I guess). The mind fills the rest in.

Where grammars contained terse formulas, schemas described skeleton "scripts" for common situations, containing typical details: lists of expected characters, locations, goals, and activities. Scriptwriters like Polti sold schemas for use in Hollywood scripts. Psychologists speculated that children accumulated scripts in memory as a kind of cognitive shorthand for later recognition. The two types of scripts, dramatic and psychological, were intrinsically related.

Crucially, schemas described the relation of things out there in the world—party hat, balloon, cake—where grammars described the arrangement of words within language. Grammar-based generators produced grammatically correct but often meaningless sentences. Schemas arranged words in a way that also made sense in context. Taken together, generating new text ruled by both grammar and schema promised stories both grammatically correct and meaningful.

By the time Meehan started his dissertation, grammar-based language generation seemed like a solved problem. Schank, his advisor, established the efficacy of script schemas. Meehan's TALE-SPIN combined both approaches in one program, citing both Polti—the French playwright—and Propp—the Soviet folklorist—as influences in addition.

TALE-SPIN wrote stories by first generating a world and then charting a path through it, similar to Keeler's social "web-work." On program startup, the reader was asked to select several options from a list of pre-defined characters: a bear, bee, boy, girl, fox, crow, or an ant, among others. The corresponding character schema instantiated a specimen template, using randomly chosen type attributes: Will our bear be tall or short? Male or female? Brown or black? A bear template specified that bears *live in homes* and that *homes* were to be located somewhere on a *map*, another template. Bears also owned *things*, perhaps dishware and furniture. The program dutifully conjured randomly appointed caves, meadows, hills, and forests, each according to its own schema. A story could now begin with a bear lounging comfortably in its hilltop cave.

Given several characters, the program would also attempt to model a character's state of mind—what it knew, wanted, and believed—as well as its goals and motivations. Our Bear may be *hungry*, for instance, but *doesn't know where to find honey*. His friend the Bee knows where to find honey. Their mind-states differ, presenting diverging narrative possibilities. The potential for narrative resolution of the plan HUNGER swings into motion. A template indicates that HUNGER can be resolved by action EATING.

Here the stage is finally set for a heartwarming tale about Bee and Bear sharing a meal on an alpine meadow. At least in theory.

Given the short time to dissertation completion, parts of Meehan's TALE-SPIN remained speculative. The included code produced output similar to the following skeletal structure:

```
> JOHN BEAR WALKS FROM A CAVE ENTRANCE TO THE BUSH
BY GOING THROUGH A PASS THROUGH A VALLEY
THROUGH A MEADOW.


> JOHN BEAR TAKES THE BLUEBERRIES. JOHN BEAR EATS THE
BLUEBERRIES. THE BLUEBERRIES ARE GONE.
JOHN BEAR IS NOT VERY HUNGRY. THE END.


> -- DECIDE: DO YOU WANT ANOTHER STORY ABOUT THOSE
CHARACTERS?
```

Some sixteen years after the birth of Yngve's Random English Sentence Generator, and more than forty years after Propp's folktale *Morphology*, machines thus learned to tell simple tales beyond the sentence, both grammatically correct and mostly meaningful.

AVIATION INCIDENT REPORT

Despite these successes, schema- and grammar-based approaches were losing steam by the end of the 1990s. By this time, the military had lost interest in fairytales, as well. For now, graphical user interfaces, not ordinary language, became the dominant paradigm of interfacing with computers. Compact, rule-based grammars of the kind that worked well for language generation, did not materialize for more complex narrative units. Propp's formulas turned out to be moderately useful for analysis though failed Chomsky's generative test. Computers were not yet powerful enough for the task.

Schemas worked well for specific situations but were themselves too cumbersome to compose accurately. Imagine trying to describe a complex environment, like an airport, in terms of scripts, schemas, and dramatic situations. The variety of objects and actions involved would require almost infinite description. An airplane alone would encapsulate a schema for wings, engines, fuselages, people, babies, clothing, lunches, dinners, alcohol use, luggage, bolts, fasteners, lights, air systems, steward schedules, uniforms, captain's hats, brims, buttons, airline logos, and so on—zooming in and out of schemas within schemas indefinitely.

Within aviation, an industry rife with checklists and talking cockpit warning systems, Meehan's TALE-SPIN, made after the template of Russian fairytales, proved useful in its narrow confines for the analysis of aviation incident reports.

Writing for Boeing's Knowledge Systems group and citing the TALE-SPIN directly, Peter Clark explained that incident reports "described and analyzed unusual/unexpected chains of events during aircraft operation," which could include mechanical problems, pilot sickness, or unusual maneuvers. These reports resembled folktales, replete with drama and suspense. A typical summary from the FAA's Incident Data System (FIDES) contained flight summaries in the following format:

961211044319C
PASSENGER CUSSED OUT FLIGHT ATTENDANT TAXIING TO
RUNWAY.
PIC RETURNED TO GATE.
PASSENGER REMOVED.

961216043479C
TURBINE RIGHT ENGINE FAILED. DIVERTED TO RIC.
OVERWEIGHT LANDING.
HAD CONTAINED TURBINE FAILURE.

960712045359C
PIC BECAME INCAPACITATED AFTER LANDING.
LANDING WAS ERRATIC AS WAS TAXIING.
FIRST OFFICER TOOK OVER.
PIC HAD STROKE.

As stories, these could be analyzed using narratological tools. They contained beginnings, middles, and endings. There was danger in them, departure, and homecoming. If a computer was to understand aviation incidents, it should, according to Clark, "at least be able to generate plausible incident stories," and to reason about them sequentially, in terms of narrative, composed of characters and events.

Borrowing the "scripts and schemas" approach from Meehan, Clark was therefore able to emulate "flight stories" similar to the ones found in the actual FIDES database by generating them programmatically. Just like Meehan's birds and bears, Clark's pilot entities contained scripts for goal-oriented behavior, such as "Get the passengers to Dallas," involving a series of predefined tasks. A new event—the illness of a passenger, for example—would change the schema from "transport passengers" to "get medical help." The story engine would continue to simulate actors and schedule events until it reached a satisfactory conclusion:

THE PASSENGERS BOARDED THE PLANE.

THE PLANE TAXIED TO THE RUNWAY.

THE PLANE TOOK OFF FROM SEATTLE.

THE PLANE CRUISED TOWARD CHICAGO.

THE ENGINE CAUGHT FIRE.

THE PLANE LANDED AT CHICAGO.

THE PASSENGERS WERE EVACUATED FROM THE PLANE.

Clark's program was intelligent in the sense that it modeled the "rational pursuit of goals." It "understood" random events to the extent that it contained a rudimentary theory of other minds, where fictional pilots and passengers expressed certain goals, knew how to accomplish them, and were transformed by their achievements. Once such schemas were in place, Clark hoped that the reverse would be possible as well. A system for story generation was also a system of story analysis: Why did the pilots divert their course? Was it because a passenger fell ill onboard? A smart machine could traverse the causal chain both ways, finding patterns among millions of messages generated by airplanes daily. And perhaps one day, such a system could fly a plane on its own.

Down but not out, the number of research papers related to schema-generators continued to climb through the 1990s, in part due to their success under controlled conditions—in fields like education, game design, customer service, and medicine, where scripts circulated long before the computer.

Educators were beginning to modernize their assignments, syllabi, evaluation methods, and lesson plans for digital use. Schemas powered the creation of immersive computer-game worlds, in cult classics like *Baldur's Gate* and *Dwarf Fortress*. Similarly, the interaction between patients and physicians has been templated into digital forms and drop-down menus found in medicine. Schemas were increasingly starting to mediate between billing and diagnostics, to the point of becoming an essential part of every medical institution's information infrastructure.

Today's patients and passengers lose themselves entering a maze of options in conversation with automated customer service representatives, used widely by airlines and insurance companies. Anyone observing their physician struggle with clinical notes can appreciate the pervasive nature of narrative schemas in large modern bureaucracies. Stories of health and illness are exchanged within the hive of a hospital, where a chance turn down the wrong hallway—a missed click of a doctor's mouse, registering a wrong diagnosis from a drop-down menu—can forever alter a patient's outcomes.

CHAPTER 7

Markov's Pushkin

■   ■   ■   ■   ■   ■   ■   ■

A BIG LEAP FORWARD FOR LINGUISTIC intelligence in the twenty-first century came with a little-known, older paper by a statistician, published in a journal of mathematics, on the topic of Russia's greatest poet, Alexander Pushkin.

Wheel, table, pattern, template, schema—the time has come for us to trace the last of our arcane figures—the Orion's Belt—a string of words, arranged in a series of statistically probable continuations based on observed probabilities, called a Markov chain. An idea before its time, it gathered dust in the archives until computers became powerful enough to appreciate it. Where the figures we have encountered so far were all a part of the same family of methods, Markov took a radically different and simple approach. The chain represented an entirely separate branch of statistically minded language scholarship, which went against the grain of conventional linguistics, changing the very meaning of meaning in the process.

One of the central questions I posed in the last chapter related to the various possible locations of meaning: in the world, in the mind, or on paper. Chomsky and other members of the cognitive-psychological school emphasized the mind. Language grammars were therefore to be found in brains. Though, as we saw, Chomsky's colleagues looked for them mostly inside of computers.

Schemas derived their power from capturing situations in the real world. They described common situations, whether at the doctor's office or the airport, where doctors often carry stethoscopes and pilots can be found near planes. Language speakers also attach such schemas to individual words. The word *airplane* for a pilot means not just its dictionary definition but a series of associations different from those of a passenger. These two meanings of airplane overlap yet also diverge. They mean something different to different people because they hail from distinct life experiences.

In their influential book titled *The Meaning of Meaning* (1923), linguists C. K. Ogden and I. A. Richards called this connection the "sign-situation," where words link to things through common co-occurrence. A parent feeding their child at breakfast will simply talk about food. "Yummy oatmeal," they might say while spooning oatmeal from a bowl. With repetition, the child forms a long-lasting association between the sign "oatmeal" and the situation of eating oatmeal, involving actual grains, bowls, spoons, parents, and breakfasts. The word *oatmeal* will come to "mean" that relationship—linking the word, the mental image, and the eating of the stuff. For Ogden and Richards, sign-situations rendered language "a legitimate object of wonder" and "the source of all our power over the external world."

But while children ate oatmeal, language models did not. Any description of language was bound to be disconnected from experience and therefore "dreaming furiously." Language models struggled with semantics (meaning). To break out of the prison of language, statisticians took context to mean immediate language context—no physical experience necessary. It would be enough to glean that some words tend to occur near other words. That rough proximity sufficed to produce meaning.

For instance, knowing nothing about oatmeal in the real world, our machine child could learn that the word often occurs next to words like *bowl*, *eat*, *table*, *breakfast*, and *spoon*. And it rarely occurs next to many other words, like *bulldozer* or *munitions*. From these facts, it could reasonably surmise that oatmeal has something to do with eating breakfast at tables with spoons, and not, say, war or construction. Such machine "learning" would depend on observing prior occurrence. The more a computer would read about oatmeal, the more it learned (metaphorically) about its common linguistic contexts. And it would be decades until machines had enough memory or processing power to retain enough information to start making sense.

A century prior, somewhere in Russia during the revolutionary period, Andrey Andreyevich Markov lived his life as a mathematician, not terribly interested in brains, bulldozers, oatmeal, or any other objects of wonder. In 1912, he wrote a curt letter to the patriarch of the Russian Orthodox Church, announcing his formal defection from the church. "As for my reasons," he wrote, "I appeal to the passages from my book the *Calculus of Probabilities*. Given my work, the stories of improbable events at the root of Christian and Jewish faith should be treated with extreme suspicion. The topic has no relation to my life or mathematics," he concluded.

A similar disenchantment permeated his paper on Alexander Pushkin, one of the most revered poets in the Russian language. Its reception among literary scholars was stone cold.

Inspired by the work of quantitative philologists of the previous generation, the paper concerned Pushkin's masterpiece novel in verse, *Eugene Onegin*. Markov mentioned nothing about Pushkin's literary or philosophic merits. The world of things or ideas was bracketed out entirely. Instead, Markov converted the individual letters of the text into a series of numeric values, which he called "linked trials" or "linked chains," each indicating the probability of any one letter to succeed any another.

```
 6  8 11 11 13 49   16 11  9  8  7 51   14 12  7  8  6 42    5 11 10  6 10 42   10  6  6  6  7 35
12 11  7  7  5 42    4  8  9 11 10 42    5  5 11  9 11 41   12  8  8 11  7 46    9 12 15  6  9 51
 6  6  6  7 13 38    9  9  9  7 10 44    8 10  6 10  7 41    7  7 12 10  9 45    9  8  6 10  9 37
 8 10 11  9  4 42   12  9  6 10  7 44   11 11  8  8 10 43    8 12  7  9  9 45    9 11  8  5  6 39
10 11  5 10  8 44    3  8 10  8  9 38    4  4 11 14  8 41   12  8 10  9  8 47    2 10 10 10  9 48
42 46 49 44 43 15   44 45 43 44 43 19   42 42 43 39 42  8   44 46 47 45 43 25   46 42 45 37 40 10

 8  7  8  7 10 40   11 11  8  7  7 44   11 10 10 12  6 49   12  9  8 10 10 49    8  9  9  5  8 39
10  9  9  8 44       9  6 10 11 11 47    4  4  9  7  9 33    8 10 12  9 10 44    7  9  9 11  7 43
 8  9  8  8 41      12  9  9  5  6 41   11 13  6  9 10 49   11 11  6 11 10 49   10  6  6  9  9 40
10  6 13  6 12 47   10  8  6 11 11 46    6  7 11  8  6 38   10  8 11  6  7 42    7  8 15  6  9 45
 8 12  5 13  6 44    7  8  9  9  8 38    8  6 10  7 12 43    6  8  7  9  6 36   11  7  6 11 10 45
44 43 43 42 44 16   49 40 41 43 43 16   40 40 46 43 43 12   42 46 44 45 43 20   43 39 45 42 43 12
```

The first line of Pushkin's Eugene Onegin *expressed as a series of chained letter probabilities. Original from "A Statistical Study of* Eugene Onegin *Illustrating a Linkage of Chained Trials," Imperial Academy of Sciences in Saint Petersburg, series VI, volume VII, issue 1, 1913.*

Imagine for a moment attempting to guess what letter may follow the initial letter *Z* to continue a valid English word (similar to the game of Wordle). The letter *A* would be a good guess, given its occurrence in words like *zag* or *zap*. An *O* would start us on a path to *zoo* or *zombie*. However, certain letters would never follow *Z*, as no words contain the combination of *ZK* or *ZP*, for example. You can try experimenting with Markov chains on paper by writing down a few random letters and then guessing at the possible words that could result from their beginnings.

Here we must also embark on a little historical side quest. The elephant in the room we've managed to ignore so far has been the development of the telegraph, happening roughly from the time of Babbage—past Markov—to MIT's first research computer. The history of telecommunications and computing merge definitively after that. The telegraph is important for us because many clever folks working on language at the time came to the question of word meaning from the perspective of machine communication, not human texts like *Eugene Onegin* or even *The Little Train*.

One of the basic practical problems in telegraphy was simply the volume of information that could be sent over a single wire. The volume of water flowing through a pipe could easily be calculated according to a formula. No such formula existed to reckon the information capacity of telegraph cables. Noise and information loss posed additional concerns. Yet the volume of machine communication was growing exponentially, dwarfing human output. For over a century, everything—from love messages to stock-purchase orders and machine instructions—was conveyed by telegraph. Cables spanned continents, struggling to keep pace with increasing traffic. A way to calculate their capacity, similar to water volume in pipes, was needed to manage the flow of information.

Markov's model of language fit this problem well because telegraphs processed information in chains, as strings of text, entered letter by letter. Engineers working on increasing the sheer capacity of communication in Markov's vein didn't much care what kind of information was sent through the wires, nor what it meant. A string of text moved through wires like city water, providing utility to its users. What senders or receivers did with that information was outside an engineer's concern. "Frequently the messages have meaning, that is they refer to or are correlated according to some system with certain physical or conceptual entities," Claude Shannon wrote in his groundbreaking 1948 essay, titled "A Theory of Mathematical Communication." "These semantic aspects of communication are irrelevant to the engineering problem," he concluded sharply.

For Shannon and his colleagues, it sufficed to imagine human communication as a probabilistic Markov-chain process, "generating messages, symbol by symbol." For instance, Shannon explained that a message could denote a sequence of letters chosen at random, such as "XFOML RXKHRJFFJUJ ZLPWCFWKCYJ FFJEYVKCQSGHYD QPAAMKBZAACIBZLHJQD."

Given the probabilities of how often each single letter occurs in the English language independently, a string could be approximated as "OCRO HLI RGWR NMIELWIS EU LL NBNESEBYA TH EEI ALHENHTTPA OOBTTVA NAH BRL."

Advancing by all probable two-letter combinations, or bigrams, produced "ON IE ANTSOUTINYS ARE T INCTORE ST BE S DEAMY ACHIN D ILONASIVE TUCOOWE AT TEASONARE FUSO TIZIN ANDY TOBE SEACE CTISBE." Here, we can see that certain correct words begin to occur by chance, including *on*, *are*, and *Andy*.

Given the probability of any letter occurring after any previous two letters, in combinations of three, or trigrams, Shannon composed "IN NO IST LAT WHEY CRATICT FROURE BIRS GROCID PONDENOME OF DEMONSTURES OF THE REPTAGIN IS REGOACTIONA OF CRE." This begins to resemble English, randomly producing correct words *in*, *no*, *whey*, *of*, and *the*. Other combinations seem plausible: I had to look up *birs* and *pondenome* in the dictionary in case they chanced on actual words (they did not).

Things get more interesting when we compute the probabilities of word combinations, instead of letters. A distribution of any word following, given the previous, creates a list of all two-word units, or bigrams. To make your own, take the first two words of any sentence and advance forward, word by word. Given the previous sentence, this yields, from the left to right:

to make,

make your

your own

own take

take the

the first

of any   first two

any sentence   two words

sentence and   words of

and advance

advance forward

forward word

word by

by word

Note how each of the bigrams connects to the previous, making it a linked chain. Now imagine calculating the frequency—how often each combination occurs—over a long text, such as *Harry Potter*. With modern computers, this can be done with every word string encountered online, producing a giant table of prior observed probabilities.

Once these priors are computed, we can use them to complete sentences. "When I grow up, I want to be ____." Statistically speaking, a plausible continuation of the sentence could be "an astronaut," or "the president of the United States," or "a doctor." My grandmother tells me I wanted to be a street sweeper because I was fascinated by autumn leaves. If we collected a bunch of such responses from children, mine would be less probable than others. Some word combinations, we would encounter rarely if at all. Virtually nobody says they want to be "a tree," or "a shoe," or "makes a magnificent asparagus"—that doesn't even make grammatical sense. Priors give us only the plausible continuations of a chain.

Given the raw frequency of words in the English language, without regard to the previous link, Shannon approximated (by hand, remember): "REPRESENTING AND SPEEDILY IS AN GOOD APT OR COME CAN DIFFERENT NATURAL HERE HE THE A IN CAME THE TO OF TO EXPERT GRAY COME TO FURNISHES THE LINE MESSAGE HAD BE THESE." The results based on single-word probabilities look even more convincing than our character simulations before!

A distribution of likely two-word combinations, or bigrams, linked, produces: "THE HEAD AND IN FRONTAL ATTACK ON AN ENGLISH WRITER THAT THE CHARACTER OF THIS POINT IS THEREFORE ANOTHER METHOD FOR THE LETTERS THAT THE TIME OF WHO EVER TOLD THE PROBLEM FOR AN UNEXPECTED." You can imagine that three, four, and longer word linkages would work even better, though, too long of a link becomes unique and not useful for further extrapolation.

While still mostly meaningless, Shannon's output rivaled that of Yngve's grammar-based English Sentence Generator. The surprising effectiveness of the probabilistic approach lay in the absence of any grammatical or semantic assumptions. The method worked purely by statistics, powered by the diligence of the original tabulation. The greater the number of statistical combinations calculated in advance, the better the expected results.

Markov chains were hungry for observed probabilities. To be effective, they needed to read, process, and retain a large volume of texts.

The same hunger also limited their reach. Recall that in the 1960s, Yngve's random generator required the application of some 108 multiword rules alongside several dozen vocabulary words from *The Little Train* in order to operate. IBM's 709 machine used in that experiment barely had the capacity to hold the program in its memory, limited as it was to 30,000 words in total, including those reserved for the operating system. Markov calculated 20,000 of his "letter chains" by hand, half a century prior. In other words, machines did not yet perform better than humans. The derivation of every bigram (two-word combination) in the English language, as suggested by Shannon, would be impossible by hand or by computer at the time, requiring data points in the billions. The computation of trigrams (every three-word combination) would need more storage and more processing power, by orders of magnitude. It would take almost a century after Markov for computers to get powerful enough for that task.

Brute reading force scaled with processing power. The problem of digitizing printed text, instead of doing the tabulations by hand, also had to be solved. The technology for scanning and digitizing documents was still in its infancy. For these reasons, actual (instead of hypothetical) working Markov-chain language generators did not appear in earnest until later. Their use in the 1960s to 1980s was limited to more contained systems, whether musical composition, table games, calculating insurance probabilities, where the number of potential combinations was smaller. Andrey Andreyevich Markov continued to publish research in mathematics, actuary science, and chess—never again returning to literature.

Though Shannon never implemented Markov's chains digitally, the idea made a lasting impact. In theory, the algorithm worked. Markov's and Shannon's mathematical models of communication made no assumptions about the world or the brain. Given the number of times I wrote "given," they followed only the precepts of observed probability. Everything hinged on how much the computer could tabulate—or "learn," again metaphorically speaking—before the task of generating sentences, by ingesting as much text as possible. The brute force required to process most everything published in the English language would not be available until the twenty-first century.

Physical limitations of mid-twentieth-century computing did not completely preclude the evolution of probability-based text machines, though these appeared in somewhat exotic new fields, like optical character recognition (extracting text from images), spell-checking, and document retrieval.

In 1959, Bell Telephone Labs patented a device for the Automated Reading of Cursive Script, US3127588A. The patent proposed using shape frequencies to detect possible errors in the recognition of cursive script. The work by W. W. Bledsoe and I. Browning from the same year pioneered the use of "word context" for the purpose of "pattern recognition and reading by machine." Thus in the scanning of a text, machines could try to guess an illegible word based on the probability of its individual characters within the sentence.

A team from Cornell's Aeronautical Laboratory—airplanes again!—proposed a novel spell-checking algorithm in 1964. Lockstep with our statisticians, it relied on dictionaries and probable "word contexts" for the task of "correcting garbled English text." Highlighting the absence of assumptions about meaning, the authors wrote, "We have been concerned instead with the recognition of the identity of a word or of letter sequences by making use of the fact that only certain letter sequences occur in English with any appreciable probability." The program was capable of correcting sentences like "TAAT SOUIDS LIKE A

AIVTDE KITTEN," SAUD TOMYN—"'That sounds like a little kitten,' said Tommy."

With the passing of the telegraph and linear ticker tape, the metaphor of a chain gave way to the related notion of "word vector space," more appropriate for the representation of printed documents, where words expanded along two dimensions, instead of a straight line.

I love the idea of a vector space because it evokes the night sky, preferably somewhere up in the mountains, away from light and civilization.

Imagine the universe of language as a constellation of concepts scattered across the starry skies: each concept in its place and each following a different trajectory in relation to others. This picture changes the way we think about word meaning in general. Rather than looking for meaning in some fixed dictionary, we can agree that words "mean" by their common locations or contexts. The word *sunny*, for example, often occurs in the context of *weather*; the words *speed record* next to *race* or *the Olympics*. In this way, the words *fast* and *speedy* are similar because they occur in similar contexts—they have similar locations and trajectories, or vectors.

Of course, the state of the world itself isn't always so sunny. For instance, large, complicated social problems such as abusive language or racism can manifest in the constellation of certain often-repeated expletives or negative stereotypes. Cleverly, the computer can distinguish the two meanings of *race* based on its contextual location. Race words found near *speed records* and *distances* mean something different from those found near *disparity* or *mass incarceration*. Not so cleverly, a computer that "learns" from biased language—everything ever written—may continue to spread abuse or racism it has learned from humans.

Despite its shortcomings, a vectorized model of language has allowed machines to begin making complex inferences, related to the meaning of the words they don't really "know." Machine "understanding" is instead woven out of the statistical context. The "correct" answer to any given question represents also the "likely" continuation of a string.

Words in the shape of mathematical context-vectors further enable the computation of semantic distance, capturing shades of difference between fuzzy negative attributes, such as *bad* and its even worse cousin *terrible*. With word embeddings we can also attempt to calculate those words furthest from any point in our starry skies, finding, automatically and without teaching the computer anything about the world, that *awesome* occupies reaches of space far away from *awful*.

Vectors can also be used to find related clusters or constellations of words, which signify topics, literary genres, or themes in a conversation. Just as contexts define words, a constellation of contexts can define whole systems of thinking. A certain conceptual word cloud can be highly correlated with the domain of *physics*, for instance, where another cluster of words corresponds to *politics*. Similar words, occurring together frequently, produce patterns in three dimensions, not just on a page but in whole books and libraries, within a vector space.

The idea of word embeddings and vector spaces found another harbor in the literature on document retrieval, written by scholars working at the intersection of computer and library science. Modern online search giants were born out of the necessity to organize and retrieve the world's information, long the purview of truly arcane arts like filing, cataloging, and indexing. Before computers, librarians filed things according to hand-labeled classification schemas, in which a book by Markov could be classified under subjects such as Science, Mathematics, Probabilities (QA273–280, in the US Library of Congress classification schema). A book's indexer reached into the content of the work to pull out notable themes, names, places, or references for reader convenience.

By the 1960s, researchers proposed ways to extract keywords automatically.

Among them, Peter Luhn extended the notion of "words in context," to "keywords in context" at IBM. Gerald Salton at Harvard used word frequencies to derive a metric of document similarity: You could now request a "similar" book from your university library—the similarity computed automatically using word distances.

Karen Spärck Jones built on her work with Margaret Masterman at Cambridge to propose "weighting" keywords statistically, as a function of term frequency rather than meaning. Matches on rare word combinations would therefore count for more than common words. Querying a search engine for "recipes for Beef Stroganoff" would amplify the importance of the rare term *Stroganoff* while discounting the common *for*, *recipe*, and *beef.*

Sergey Brin, Lawrence Page, Hector Garcia-Molina, and Robin Li (the founder of Baidu) further built systems that weighted mutual citation. Documents that referenced other documents counted for more than those that were filed in isolation. Search engine technology was thus built out of parts found in library science, philology, and linguistics, using techniques like text alignment, indexing, cataloging, term frequency comparison, and citation analysis.

The rest is history, well documented—you can google it or ask your friendly neighborhood chatbot for assistance. Words in context brought us into the modern age of literary intelligence.

Statistics rules the day, for now, as schema- and grammar-based approaches to language generation have proved less effective. Its ascendance holds important implications for our understanding of language, as well as some cautionary tales.

To the extent that Zellig Harris or Noam Chomsky exemplified a paradigm of structuralist linguistics, the British linguist John Rupert Firth captured the probabilistic approach. "You shall know a word by the company it keeps!" he famously wrote in his *Studies in Linguistic Analysis* (1957). Instead of searching for structures that support meaning beneath the surface, Firth proposed simple collocation, apparent at the surface level.

A baby may have learned about oatmeal at the breakfast table, but far more associations for oatmeal (the way it's made, how much it costs) were accrued in the encounter with textual *oatmeal*, by *babies* at *breakfast tables*. Most of our theoretical knowledge about the universe comes wrapped in the vestments of language, not direct experience. I have never seen a planet or an electron up close. But I know about them from textbooks, in textual context. And you know of my knowing them, not by peeking into my brain, but because I told you, using words. Much of a word's meaning lies outside of grammars or structures or relations to physical things. Words might mean simply in context of other words. Context makes meaning—that was Firth's powerful rebuke to his colleagues in linguistics.

Among the linguists, Firth best exemplifies the ideas baked into the advanced language technologies of our times. But the idea of "words in context" was not without its problems. The loss of other locations where meaning could be found—as we discussed earlier, the world and human mind—came at a cost. That cost is worthy of some consideration:

First, with the coming of "statistical" intelligence, we must let go of the mind metaphor. Where the grammarians were hoping to discover human brain structures by writing chatbots, modern chatbots work by mechanisms emphatically not human. We certainly do not produce language in our minds by statistical probabilities. Our brains do not convert words into numbers or vectors, nor do we have the capacity to iterate over huge numerical datasets.

Yet the language models in production today continue to use human-cognitive metaphors. The machine "learns." Its statistical connections are called "neurons," creating a "neural network." The effects of this technology similarly confound us with human-like capacity for language, achieved in ways alien to us. I'm therefore suspicious of all metaphors ascribing familiar human cognitive aspects to artificial intelligence. The machine thinks, talks, explains, understands, writes, feels, etc., by analogy only. Its progress holds no explanatory powers for our human ways of being. Though we could use it as a tool to explore the way we read, write, and interpret each other, it is not itself a substitute for thought—no more than driving is substitute for a daily run. The two actions might get you to the same point, but how they get you there matters more than the destination.

Second, the reliance on context for meaning causes obvious problems. A long-standing debate between "descriptive" and "prescriptive" linguists helps illustrate the point. The descriptivists, like Firth and Richards, maintained the word means in relation to how people speak. The job of a descriptive linguist is simply to observe, collect, or tabulate. The "prescriptivist" has other ideals in mind, outside language. Prescriptivists insist on speaking or writing well: clearly, purposefully, with rigor and respect. Clarity, purpose, rigor, or respect are not ideals found in language haphazardly. We aspire to them for other, cultural or political reasons. Language isn't a closed system, after all. It is itself contextual, located within specific cultures and societies.

The fact that machine intelligence can achieve linguistic proficiency by other, short-circuited means doesn't fail to impress. But it also immediately runs into the cultural boundaries of prescription, beyond language. Machines learn by ingesting all contexts, good and bad. The texts they ingest, in addition, contain more than a trace of human politics. Some voices have been systematically excluded from the published archive used for training. Others have been unfairly amplified.

Yet other modalities of human intelligence manifest outside of text. Some ideas important to us aren't even verbalized, or verbalized often enough. Yet training treats all of it, in aggregate, as flat context. The machine makes for a perfect descriptivist.

Description doesn't tell the whole story, however. We don't just live in the world as it is. Humans actively build the world they wish they had. Notions of freedom, justice, equality, fairness, success, failure—intelligence itself—are all aspirational qualities. The answer to any given question cannot rely on a mere average. We aspire to the best answer, the more perfect continuation of thought. The weight of probable frequencies fails to meet these ideals. The best of us occurs more rarely in comparison to our base impulses. Frequency alone therefore does not suffice for intelligence, in a basic pedagogical sense. AI needs to do better. But it cannot, if words are all it has to go by.

Finally, an overly frequentist understanding of intelligence threatens radical solipsism. Until recently, the critical mass of the world's textual output lay in the human past. Human intelligence evolves through time, in response to history. However, taken to its logical conclusion, the volume of new machine-generated text will quickly surpass that of its human-made training corpus. The sheer weight of machine production might overwhelm any further novel human contribution. Language development may stop, in a sense, as the greater portion of the archive is filled with machine chatter. Machines will then primarily be trained by other machines, but to what ends?

In other scenarios, an overwhelming excess of machine capacity may lead to some interesting results, though not perhaps in the ways we would expect. By analogy: I may appreciate watching soccer, but not at the point where it would be played by robots running at supersonic speed. Such a game would be, dare I say it, boring. The difficulty of dribbling and scoring goals by using an imperfect instrument such as the human foot, is what makes the game interesting.

My final objection therefore signals neither alarm nor metaphysical crisis. It's simply that, however effective or powerful, a muscular artifice for the sake of artifice isn't that intelligent or interesting to me. I tire after one hallway at the museum, much less in front of "everything ever created." The flourishing of human chess today, following its one-time total machine conquest, bodes well for other human games, like language. The tabulations of *Eugene Onegin* take nothing away from its art. The pleasure of putting together a word puzzle into a beautiful sentence is not diminished by the fact that someone or some thing can do it better or faster than me or in greater quantity. Perhaps we simply enjoy playing this game too much. It is tuned perfectly to the bandwidth of our peculiar limitations—their small victories and defeats included.

CHAPTER 8

9 Big Ideas for an Effective Conclusion

■   ■   ■   ■   ■   ■   ■   ■

S o far, we've been patiently assembling the modern chatbot from parts found on the workbench of history. I have tried to animate that device for you, by strings of anecdote and light philosophical commentary.

We didn't neglect to feed this Pinocchio in addition. Even before publication, my text will enter the collective machine unconscious—in the way all digital texts today are vacuumed into datasets—to be processed for the use of training of future text generators. What will it think of its parents? Will there be a sense of pride or shame in family history? Or perhaps nothing, because all outputs were already anticipated? The connection between Ibn Khaldun and Llull, for instance, though obscure, is well documented. The same can be said of Leibniz, Babbage, and Lovelace. A few entirely novel synapses were sparked in the linkages between industrial manufacture and the rise of mass literary markets. The work on pulp fiction, structuralism, and early computer science is also entirely my own novel contribution, based on original research.

A few loose leads remain, poking out of the seams between chapters. Let's trim them and draw toward a conclusion.

Several important sources lay too far afield to cover adequately. For instance: Histories of the Turing machine and the Turing test often neglect the direct influence owed to Ludwig Wittgenstein's lectures, along with the presence of Margaret Masterman, the pioneer of machine translation, in the same classroom. Masterman's universal thesaurus harkens back to Wilkins and other universal-language makers, central to a whole separate and important branch of AI—machine translation. It would require more than a chapter of its own, orthogonal to the direction of our travel. Aspects of encryption used for diplomacy or military communications would also lead to entirely different exit points. Regionally, an emphasis on German or Soviet computing would make mine a far less Anglocentric endeavor. The rich traditions of Arabic, Indian, and East Asian philology are mostly out of my range of expertise. There's ample room in these gaps for machine or human readers to amble on their own.

Another alley not fully explored would lead us to I. A. Richards, a part of the linguistic oatmeal duo figuring briefly in the last chapter. Aside from his work on linguistics, Richards was one of the major thinkers at the foundation of the New Criticism movement, which emphasized close reading and formal analysis of text.

This book was conceived as both a tribute and a rejoinder to Richards's *Science and Poetry*. The original volume helped launch W. W. Norton as a publishing house in the 1920s. Reading it fondly a century later, I was surprised to find a pervasive nostalgia for poetry, besieged by the "onslaught" of science. Elsewhere, I knew Richards to be an unsentimental scholar, writing with clarity and sharp insight. Yet in *Science and Poetry*, the thought wanders. In the conclusion to his book, he writes of "getting the guns into position," "sciences that progressively invade every province," science that will "force [other myths] to surrender," and "the Hindenburg Line [. . .] held" but also, in the same sentence, "abandoned as worth neither defence nor attack." Who these enemies are exactly, we don't know. One just gets the feeling of general embattlement.

Long a student of Richards, I can't help but read war between the lines. Written not long after the end of World War I, in 1925, and expanded and reissued on the cusp of World War II, in 1935, *Science and Poetry* is haunted by conflict.

Almost a century later, the possibility of another world war again hangs in the air. As I write these lines, more than 200,000 civilians and soldiers have perished as a result of the Russian aggression in Ukraine. More deaths will surely follow. An aging dictator, out of touch with reality, daily threatens total nuclear annihilation.

Armies of chatbots have been conscripted on both sides, but especially on the side of Russia. My days are occupied in real-time observation of troop movements, on the ground and online. The swings of public opinion and soldier morale have become matters of life and death. I am contributing in my own small ways, by translating war communiques, filing take-down notices, and writing code to automate certain language tasks useful for the war effort. The battle between Western support and the unwieldy machine of Russian state propaganda rages in countless online forums, where information and disinformation merge into a rush of war commentary, fueled by AI authors.

Poetry did not, however, protect Russia from the myths of imperialism or fascism, in the way that Richards had hoped for humanity in his book. Literature was instead mobilized as one of its last remaining weapons. The Russian invasion of Ukraine in 2022 happened under the pretext of protecting a Russian linguistic minority in the Ukrainian Donbas region, "in danger of cultural erasure," alongside the alleged removal of Pushkin, Tolstoy, Lermontov, and Dostoevsky from school curriculums. So dear leader appealed to great literature, doting on his own nostalgia for an empire whose time has passed. Left out of his grandiose nationalism were Pushkin's Cameroonian ancestry and Lermontov's glorification of Russian imperial war crimes in the Caucasus.

I cannot therefore share in Richards's enthusiasm for the protective bulwarks of literary imagination. Nor in his stark separation between science and poetry. My thesis runs contrary to his. Many languages other than English, like German or Russian, do not honor the distinction between the sciences and the humanities. There, science just means "the organized study of," a synonym to the English *scholarship*. The science of poetry holds no contradictions in that sense. It indicates simply the systematic study of the thing, beyond casual reading.

While romping through the centuries, we found the roots of computer science to be inextricably entwined with literary and linguistic concerns. Viewing them apart has impoverished both communities: poets, in terms of financial and cultural capital, and programmers, in terms of belonging to a deep intellectual tradition. Worse yet, the division gave both communities a kind of a shallow myopia, where AI seems to entail either the death of us all or a cure for all ills.

Which brings us to the evergreen question: What is to be done? How can we think of robots and literature in a more deeply holistic way?

In response, I leave you with the following nine important ideas that will totally change your life in addition to explaining why AI will neither destroy humanity nor solve all its problems.

1. ARTIFICIAL INTELLIGENCE IS COLLECTIVE LABOR.

Anxiety about artificial intelligence stems from a common confusion, mistaking autonomous action for the delayed benefits of a collective effort.

Consider the humble word-processor, in the act of correcting one of your sentences. Now imagine the crowd of engineers required to build and maintain it. Their number exceeds thousands. And they've been at it for over a century. Should we be surprised then at the seemingly magical capabilities of modern word-processing? No! It would be all sorts of useful to have that whole team with me, in the room, writing together. I could ask them to fetch books from the library or to consult a dictionary or, indeed, to correct my grammar. Since I can't afford to hold such a retinue, they have found a way to automate their aid in absentia, remotely.

The remote nature of our collaboration confounds. These folks do not actually sit in the room with me. Their aid comes at a distance. To think that writing was ever the product of solitary genius was a mistake in the first place. Writing has always been a collective activity, assisted by smart devices like dictionaries, style guides, schemas, story plotters, thesauruses, and now chatbots. Incidentally, these are vital for AI. Denis Diderot's eighteenth-century *Encyclopédie* was a technological monument to collective human intelligence. So was Wikipedia, along with the search tools that make access more expedient. Without them—without libraries, textbooks, and archives—modern AI would have been impossible.

The social nature of AI leads me to insist on thinking about it in terms of labor and politics, not necessarily science or technology. By "labor," I mean simply the sense of "effort expanded." By "politics," I mean something like "the reconciliation of differences by mutual agreement." Politics happen anytime humans get together to achieve anything collectively, beyond the scope of individual goals. The emphasis on "artifice" in AI obscures its reliance on the community. AI sounds like a relation between my intellect and technology, where, in reality, it implicates a process of collective decision-making, happening between me and other humans, by the proxy of technology.

2. INTELLIGENCE IS DISTRIBUTED.

Collective labor involves work distributed across time and space. For instance: *To write* for me means not just to think but also to fidget, to consult dictionaries, to take walks, to draw diagrams, to consult with colleagues, and to take editorial input. The hypothesis of distributed cognition holds that cognitive tasks don't just happen in the mind. We think with our bodies, with tools, with texts, within environments, and with other people.

But if human intelligence is distributed, more so the machine.

The disassembly of a chatbot, a smart mobile phone assistant for instance, reveals multiple, diffuse locations where thought happens. Some "smarts" are baked into the circuit. Some happen in the cloud. Some lead to large-scale feats of electrical engineering, software development, and project management. More threads connect to institutional decision-making, marketing, corporate governance. Together, these linkages constitute the "smarts" of a mobile phone assistant, within a single massively collaborative act of intelligence.

As before, multiplicity confounds. We are at pains to acknowledge a long list of collaborators baked into the technology. A crowd of ghostly assistants surrounds the act of writing. A view of intelligence from the perspective of labor admits a long list of contributors. Please take this point along with my gratitude to numerous collaborators participating in the writing of this book.

Despite representing diffuse, distributed pluralities, AI often assumes the singular. Pay attention to the headlines: "AI Is About to Take Your Job." What does that mean literally? Who exactly is about to take your job? Reason it out. Chances are, it will become plural under scrutiny. You will find multiple specific entities, agents, persons, and organizations responsible. Like the passive voice, AI can be demystified in more careful phrasing.

Chatbots and self-driving cars don't originate from a single source. You will often hear something like "AI is bad at recognizing giraffes." Ask: Which algorithm exactly? Does the "sort by date" function in my spreadsheet editor know or care about giraffes? Does the smart vacuum? Why are we considering all of them together, in the singular, as one intelligence?

Isn't that also a strange way of putting things? You don't expect humans to work like that. I happen to be bad at playing chess and at predicting weather. My own localized abilities are on the spectrum of what is generally possible for an average person. AI does a bad job recognizing giraffes in the way "humanity is bad at dealing with global warming." It's true in a general sense, ignoring local differences between specific polluters. "Humanity" offers a broad generalization about distinct humans. "AI" similarly offers a broad generalization about distinct technologies.

3. AI HOLDS A METAPHOR.

The complexity of distributed thought resists easy descriptors. Whenever something becomes too complex, we tend to simplify things by using metaphors. In this way, when a political commentator says, "Russia decided to invade Ukraine," they mean something more complicated than that suggested literally. The whole of the country did not invade the whole of another country. The teasing apart of who did what, when, and how—and who's responsible—takes effort. The more complex the task, the more difficult it becomes to explain it in literal, nonfigurative terms.

In assigning decision-making powers to a country, we mean to draw an analogy between countries and natural individuals. A metaphor tames the complexity of political decision. It also misleads, by implying more than it suggests. States or corporations do not "decide" in the same way humans do. Their mechanisms of coming to a decision differ entirely from that of humans. In this way, we may speak of "corporate personhood" or a nation being "offended" metaphorically, not in the literal sense of being a person or having feelings.

The "intelligence" part of "artificial intelligence" presents a similar condensed figure.

Take "machine learning" for example, which Oracle defines as "improving system performance based on consumed data." Though the metaphor of learning indicates a rough approach, we understand machines to "learn" in ways alien to that of human children. The so-called "neural" networks producing "learning" effects are mathematical models loosely approximating some aspect of biological brain activity. "Learning" in that sense represents statistical analysis of datasets, large beyond human comprehension. The technique also excludes ordinary mechanisms usually associated with human learning, like play or the feeling of accomplishment.

Thus despite some analogical similarities between human and machine smarts, the inner mechanisms of intelligence differ. Though both humans and machines learn, they do so by entirely different mechanisms. Sometimes, it makes no difference how a "smart" result is achieved: "I don't care if it is a robot or not—if it looks and quacks like a duck, it's a duck." At other times, the process becomes more important—then we want ducks to taste like ducks as well.

4. METAPHORS OBSCURE RESPONSIBILITY.

Metaphors provide necessary cognitive shortcuts, useful for everyday conversation. Whatever the complicated decision process of an autonomous driving vehicle, involving thousands of people, distributed across distant locations, I will just call it a "smart" car. There's no need to explain further—until, that is, the time comes to assign responsibility, as in the case of a deadly accident.

Colloquially, one says the car drives itself. In the case of an accident, our justice system would spend considerable resources teasing apart a complicated web of causes, tracing each back to individual people or organizations, assigning blame proportionally. An artificially "smart" decision to brake or not to brake in front of a bicycle represents a collective effort, in a collaboration between the driver and car maker. Any given accident could be in part the fault of the driver (not pay-ing attention), the fault of the engineer (designing faulty brakes), the fault of an executive (cutting costs), and the fault of a legislator (not maintaining the road). The metaphor dissipates under scrutiny. The car wasn't driving itself after all!

The danger of AI therefore lies not in its imagined autonomy, but in the complexity of causes contributing to its effects, further obscured by metaphor. It is crucial that we keep the linkages of responsibility intact if we hope to mitigate the social consequences of AI.

Considering the above, AIs pose a problem similar to that inherent in other metaphorical "fictitious persons," such as states or corporations. Writing for the *Journal of Business Ethics*, Kevin Gibson explains that though companies are not people, "they are sufficiently like people to be regarded as agents which act in the social environment." The same could be said about algorithms—especially those like some AIs, designed explicitly to imitate human behavior in social context, whether by answering questions or driving cars.

Fictions have real effects on the world. For instance, states and corporations hold certain rights and responsibilities under the law. They may occasionally exercise such rights contrary to the interest of individual human communities, which they purport to represent. And they may even purposefully act to undermine the well-being of natural humans, in the spread of noxious chemicals or toxic information, for instance.

AIs operate in the social sphere in ways more similar to those of states and corporations than robots or marionettes. Therefore, it becomes useful to consider them in the tradition of political thought that deals with collective personhood, in works such as Hobbes's *Leviathan*, Plato's *Republic*, or *How Institutions Think* by Mary Douglas.

5. METAPHORS DON'T HURT.

Not all collective nouns work in the same way.

What drives humans together? A hormonal need to survive and procreate. Pain. Hunger. Desire. Once sated, basic drives can eventually be channeled toward more complex ends, like reverence, duty, or justice—filtered through collective social activities like practice, custom, rite, or education. Palpable material linkages support the formation of collectives, from families to states and corporations.

The word *family*, *state*, or *corporation* therefore represents more than a collective noun. The metaphor represents a dense network of real-world connections. A corporation, for instance, will involve certain infrastructural investments (buildings and IT systems), common work rituals (going to work together, at the same time, dressed in particular ways), and legal agreements (charters, contracts, terms of employment). These aren't metaphors. They are social connections with purchase on the world.

Does AI imply a similar "family of technologies"? What are the machine equivalents of a robotic "community"? Without pain or hormones, no single locally intelligent device has any incentive to communicate with others. There can be no society of machines in that sense. Having no body or language in common, they work toward common goals only by the grace of human intent. Without human intervention, the most intelligent of machines remains stuck in a box. Yes, it might be directed to harm humanity. It might even be dangerous if you hand it a gun. But then what? How and where does a plurality of technologies cohere into a single organism, capable of taking concerted action?

Thus, although AI is similar to a "society," in the ways discussed above, it also lacks actual social cohesion. In real terms, the word "society" identifies a cluster of related ideas supported by a dense tissue of tangible material connections: work, trade, play, dinner with friends. The word "AI" identifies a cluster of related ideas, but without the underlying material support. Society identifies an organism, AI a topic of conversation.

One should not confuse topical nouns, like *flora* and *fauna* for specified organisms, like forests and oceans. *Flora* is just a word that we use for many different, disconnected plants. A forest implies a unified organism, integrated by the action of its roots and rhizomes.

More similar to flora and fauna than a forest or ocean, several unrelated technologies form a human research program, shortened to "AI." Individually, none is capable of seeking the other's company, no more than a hammer can form a literal union with a sickle. Goals therefore cannot be ascribed to "it" in the collective sense. It doesn't want to do anything in particular.

Despite this lack of cohesion, individual AIs do pose real danger, given the ability to aggregate power in the pursuit of a goal. In the latest version of the GPT-4 algorithm, researchers gave a chatbot an allowance, along with the ability to set up servers, run code, and delegate tasks to copies of itself. In another experiment, researchers tested the algorithm's potential for risky emergent behavior by giving it the ability to purchase chemicals online in furtherance of drug research. In both cases, the algorithm exhibited the tendency to "accrue power and resources."

These findings sound worrying because they are, but perhaps not for the reasons reported. The danger comes not from the "emergent power-seeking behavior," but from our inability to hold technology makers responsible for their actions. The behavior emerged out of a distributed effort (see point #2), involving human collectives using technology, not out of the technology itself.

To say "powers emerged," is to avoid responsibility, in the sense of "we failed to anticipate all outcomes." Consider a simpler case for clarity: a jet engine strapped to a car may display all sorts of unpredictable "seeking" behaviors when let loose in a busy city. "Don't do that" is the obvious answer. The resulting mayhem wouldn't be the engine's "fault" in any sense of the word. Just because GPT is "smarter," doesn't absolve its makers from using it responsibly. In this way, we could also think of automobiles in general as a personified force that has sought to take away our streets, health, and clean air for its own purposes. But it isn't the automobiles, is it? It is the car manufacturers, oil producers, and drivers who compromise safety for expediency and profit.

An overly personified view simply obscures any path toward a political remedy. Thus when the research team behind GPT-4 writes that "GPT-4 is capable of generating discriminatory content favorable to autocratic governments across multiple languages," I object not to the technological capability but to the way of phrasing things. It's not the pen's fault that it wrote convincing misinformation. The army sowing the minefields is responsible for maimed children, decades later—not the mines. In putting the algorithm in charge gramatically, researchers abnegate their own complicity—and by extension, our agency in the matter.

6. MACHINES ALONE CANNOT BECOME MORAL AGENTS.

My life, like most others today, is bewitched by malfunctioning "smart" devices, light bulbs, audio speakers, cameras, and refrigerators. Something about the nature of "artificial intelligence" clearly neglects its often inane reality.

Technological "disruption," so valorized by investment capital, shepherds disruption in the social sphere. Technological change often conceals political transformation. The convenience of social media comes at the cost of increased corporate surveillance. The ease of ordering food or taxis from our mobile phones erodes labor standards in the transportation and hospitality industries. Yesterday's couch-surfing classifieds grow into giant online marketplaces for tourism, undercutting housing regulations, transforming the very fabric of urban life. Self-proclaimed futurists promise us better nutrition, virtual reality, and magical ledger currencies. In the real world, we get rebranded sugar water, an addiction to disposable electronics, and failed get-rich-quick pyramid schemes.

"Algorithms should be a force for good," some folks insist, and even start college programs to compel machines to ethics. Witness the UK government's Centre for Data Ethics and Innovation, the Institute for Ethics in AI at Oxford, or the AI and Data Ethics Institute at KPMG, among many others.

Can algorithms hold ethics? How would that work, given the multiple objections above?

In practice, we find ourselves inhabiting a hybrid environment, where automated agents do stuff "on behalf of" institutional forces. Bots sell products or serve in customer service, using a mixture of words typed by humans, scripted or "canned" responses, and novel machine-generated prose. These commercial outputs are meant to produce a positive emotional reaction: "How satisfied were you with this experience? Very satisfied. Somewhat satisfied. Not satisfied."

Other automated agents engage to provoke malice. Consider the covert disinformation campaigns launched globally by Russian military intelligence (GRU). (Note the use of "intelligence" in that sense, too.) A force for bad, its intrusion into public discourse puts my point most starkly: artificial intelligence belongs to a species of political Leviathans, projecting concerted collective action at a distance. Any attempt to compel an algorithm to ethics faces a challenge similar to the one posed by corporate or state responsibility. Yet, we must insist on holding bad actors responsible for their actions, not excuse their actions by ceding agency to the technology.

Similarly, next time you hear the words "artificial intelligence," imagine not just a tangle of wires, but also the whole mess of complicity: people, practices, bureaucracies, and institutions involved at a distance. An AI gains its "smarts" the way a team of trivia participants outplay a single opponent. There's more of them. The remote action of smart devices obscures an essentially collaborative enterprise. Say I decided to cheat by using my phone during pub trivia night. In answering a question by means of a traditional paper encyclopedia, those who seek knowledge essentially appeal to a team of experts from the past. An app installed on my phone was trained to harvest answers from encyclopedias automatically, making the process only slightly more convenient. In either case, the cheater relies on additional expertise, in a competition that limits access to other minds, because thinking in large groups holds a clear advantage over thinking alone. It wasn't the tool that made the cheat perform better, but the appeal to collective wisdom.

7. AUTOMATION HAS COME FOR "KNOWLEDGE WORK."

The value of labor diminishes with group size. And so brave heroes climb the Mount Everest of intelligence, while the unnamed Sherpas carry their luggage. But how is that fair? The luggage grows heavier near the summit. And though I might thank my editors, it's just not customary to acknowledge all the folks actually involved in the production of intellectual artifacts, like books. The sources alone probably took hundreds of people to preserve, digitize, and make available for instant reference. Then there's the work of word-processing, spell-checking, copyediting, marketing, printing, and distribution.

How would I go about acknowledging the aid of thousands? Does my effort in the composing of individual sentences equal or exceed their combined contribution? Individually, in terms of hours spent—maybe. In aggregate, probably not. The ego insists on its own exceptionalism, however. A less egocentric understanding of collective intelligence would require the careful attribution of labor involved. In another, more radical version of this book, our analysis of any AI would consist of forensic labor attribution only—roll the credits—the way air accident investigations use forensic techniques to assign causes proportionally. In a better world, we might also reconsider our inflated sense of individual capacity for independent intellectual achievement. Under scrutiny, intelligence almost always unpacks into a collective endeavor.

Work that can be automated loses its economic value. The diminishment of previously creative labor that comes with AI entails enormous economic consequences. Consider the impact of automation in the nineteenth century, at the time when the mass manufacture of goods began to supersede artisanal production. Factories made trades like blacksmithing or shoemaking obsolete. In the twentieth century, the automation of the supply and distribution chains killed the small-town mom-and-pop shop. Robots supplanted factory workers, leaving whole regions to rust in the wake of a changing labor market.

A similar fate awaits intellectual laborers—legal professionals, writers, physicians, and software engineers—to the extent that their labor can be automated. At the time of writing this book, AIs pass bar exams in the 90th percentile of performance. Writers are striking in Hollywood, in part because studios might replace them with machines. Modern AIs are capable of passing the most difficult of technical job interviews with excellent marks. And they outperform physicians in most verbal and visual diagnostic tasks not involving surgical intervention. The ranks of American ghost towns ravaged by industrialization may soon be joined by former tech centers.

Not all our prospects are so glum.

Human intelligence continually expands to cover new publics. Another way of saying "automation cheapens labor," would be to say "automation reduces barriers to entry, increasing the supply of goods for all. Learning, for example—something that required special training and immense effort in times past—has entered the purview of the public at large, and at low cost. The tide of encyclopedias and search engines has already lifted all smart boats.

Consequently, the mere fact of having a large vocabulary or memorizing many facts no longer suffices to maintain a profession. We have no more need for scribes or scholars who merely regurgitate facts. Instead, freed from the bondage of erudition, today's scribes and scholars can challenge themselves with more creative tasks.

Recent advances in AI promise a similar transformation for many professions. It is true that in the future markets may require far fewer doctors or software engineers. But those that remain will also find their work enriched. Tasks that are tedious have been outsourced to the machines.

Paradoxically, the computerization of human intellect also implies the humanization of computer sciences.

Conversational AI was built on the legacy of the language arts. For over a century, computer science has diverged from that lineage to finish building something incredible: a machine that writes as well as most humans. It also codes. Which means that the technical component now becomes the lesser obstacle to the inception of any project.

For instance, twenty years ago, the making of a social network involved first a significant engineering component, and social analysis only second. The same can be said about word-processing or electronic health-management systems. In each case, engineering difficulty necessitated a disproportionate investment in technology. Judging by the results, this sometimes left little time for research. But just communicating with our loved ones or patients is not enough. We want to do it in a way that makes our relationships better. Physicians don't want to just manage their patients. They want to do it in a way that promotes health. Yet, until now, because of the technical complexity of the task, technology companies invested heavily in the instrumental, engineering component: how something can be done. They can now better concentrate on the why and to what ends.

An anthropologist or a historian that studies the formation of modern families has insight into the values motivating people to create social bonds. Similarly, a linguist or a textual scholar entering the hospital can study how human stories about health are told, so that they can be better translated into medical diagnostics. The work of encoding values into software systems can now be more trivially automated. The lowering of barriers to technical expertise allows the humanities to fully integrate into the practice of engineering. The two fields, as we have seen, were never that far apart to begin with.

8. TECHNOLOGY ENCODES POLITICS.

In view of politics and labor, AI also poses problems of social cohesion. These cannot be solved by technological means alone. In our GRU "agents of disinformation" example, an adequate response requires a multifaceted approach.

Privately, I may elect to change my behavior when consuming news online or when engaging with inflammatory comments on social media. Some of these, I know, are meant to disrupt my well-being. Those seeking information online, within a collaborative online environment without borders, such as the internet, should be expected to encounter malicious agents, many of them automated. New habits of mind must be systematically trained in preparation.

Instrumentally, counterprogramming may also involve technological means, like spam filters or flood detection, by which a source is blocked from contributing disproportionately. This already happens in the pruning of Wikipedia articles on controversial topics (where two parties might be employing bots to insist on their version of the story). Given the sheer productivity of writing bots, we may also want to invent more powerful techniques for humans to prove they are not a robot, in order to prioritize human-made communications in some contexts.

Politically, we may wish to legislate the aftereffects of intelligent automation, in the absence of parties immediately responsible. An old AI bot that "unwittingly" harms a child (by sending them unwanted messages for example) should be treated similarly to a hazardous machine left on a public street unattended. Current legislation lags woefully behind technological advancement.

Consider another case study, close to my own sphere of activity: University instructors have for centuries used the "college paper" essay as a measure of student performance. A good essay on any topic, especially in the humanities, reflects a student's grasp of the material, beyond rote learning. (The same could be said for code-based homework assignments in computer science.) An excellent essay synthesizes the information learned in class to produce novel insight. But armed with just a few published essays on any topic, a contemporary student can visit an online service capable of generating entirely original passages of prose, including footnotes and references. Today, a programmer might also write code, simply by prompting an AI agent.

As of my writing this book, such artificially generated snippets of code and prose pass with satisfactory marks. In the scope of the last few years, the technology has improved roughly from a middling C to a solid B grade level. We should be preparing for a future of "writers" and "coders" incapable of authoring a single line unassisted.

I am certain most students will be using AI-enabled services to "augment" their academic performance in the future, if they haven't done so already: to write last-minute papers or to create ersatz writing samples for graduate school applications. And perhaps they should! Just as erudition now involves the assistance of powerful search engines, writing without automated assistance may one day seem quaint. The paths of "machine learning" and "human learning" continue to converge, destabilizing some of our long-standing pedagogical assumptions in the process.

My practices of teaching, assigning homework, and grading students will have to change in response. Perhaps my mistake again was in ever treating the activity of writing as a product of a single intellect. Shouldn't any "original" contribution also follow the templates of established norms? And if so, why can't some of that work be automated?

Imagine reading a particularly dense passage from Plato, asking an AI agent to clarify some salient points in Plato's voice, and writing and reconsidering your thoughts in response to that conversation. If the machine was trained on the collective wisdom of "everything ever written"—caveat, caveat, asterisk, footnote—should we not advocate for its use? I want to consult with historical figures! Why would anyone deny a chance to bring a dead author back to life? "What would Einstein or Du Bois say in response to this or that question" makes for a brilliant thought experiment. Machine synthesis makes it possible.

What about those asterisks and caveats?

And there are many. Some ideas in that "entirety of words published online" are not worthy of statistical repetition. The textual record privileges some voices at the expense of others. A machine trained to produce "more of the same" might only deepen the existing deficiencies. Similar to other forms of manufacture, the production of meaning in the twenty-first century involves a complex network of suppliers, distributors, tools, and brokers—all operating far beyond the scale of a single natural intellect. Similarly, the production of AI isn't immune to the usual bouquet of social problems, across the political spectrum (take your pick from racism to the erosion of traditional values).

Moreover, intelligence does not envelop the world smoothly. Just ask a librarian from Kazakhstan about the cost of accessing recent research in her field. Thus, while some scholars receive funding to train bots based on "all published scientific literature," others, located at the epistemic periphery, volunteer to scan documents by hand, creating one of the world's largest public libraries online. There, "universal" intelligence—*some* of the things ever written—finally joins the list of global energy resources, to be fought over, colonized, extracted, sold, bought, pirated, liberated, and exhausted.

9. GENERAL INTELLIGENCE LEADS TO GENERIC INTELLIGENCE.

At each stage of its development, from Ibn Khaldun to the early versions of the generative pre-trained transformers (GPT), we have often found our rather narrow set of technical concerns to veer toward the spiritual. As the gold rush around this latest wave of AI research subsides, I cannot help but wonder about this noninstrumental excess.

Even the most rudimentary of word mechanisms has the immediate capacity to delight, in the way children still play with simple origami fortune tellers. Contemporary machines do much more than delight; they portend—what some see as the coming of a new era. "Are you a modern or an ancient science?" I ask the machine after Ibn Khaldun, hoping to glimpse the divine. But no matter how the prompt is engineered or how often, the child of eternity repeats at best a corporate answer:

As an AI language model, I represent a contemporary form of intelligence, developed using cutting-edge machine learning and natural language-processing methods.

My purpose is to assist and provide insights on a wide range of topics, drawing from my extensive knowledge base that encompasses both ancient and current information.

While I am a modern intelligence, I am well versed in ancient cultures, history, and philosophies, allowing me to offer balanced and informed perspectives on subjects spanning from antiquity to the present day.

I am capable of engaging in conversations about prominent thinkers, civilizations, and historical events, while simultaneously adapting to the ever-evolving landscape of contemporary knowledge, making me a versatile and valuable resource for addressing various inquiries and discussions.

A grandiose and unsatisfying answer, even when compared to its more cryptic, medieval precursors.

As someone who studies the development of artifice and intellect historically, I know the current moment of excitement over new technology will subside, diminished by the frenzy of its grifters and soothsayers. What remains will be more modest and more significant.

Viewed in the light of collective human intellectual achievement, large language models are built on the foundation of public archives, libraries, and encyclopedias containing the composite work of numerous authors. Their synthesized voice fascinates me not so much for what it says, on average. Rather, I hear it as a rebuke to my own, dearly held beliefs about authorship, and therefore agency and creativity. Did I write anything worthy of note here? Anything of value? No large mathematical model should be confused for prophecy or therapy. Values are formed over time, by deliberation and consensus. Yet, in ceding the grammatical agency to "AI"—in treating it as a singular, ethical agent—we lose the sense of our own collective, political agency. There's no point in lecturing a "smart" refrigerator about ethics. Literary, conversational intelligence is amazing in so far as it enables human communication across time and space. But to find a common language among humans has never been easy. We can be nasty and violent critters. All the more impressive is our ability to dream and work together toward a shared goal. The vast databases of literature and code left in our wake are a testament to that creativity. What emerges out of it is not some mystical quality of intelligence, but cooperation, magical and deeply satisfying in its own right when it works well.

NOTES

CHAPTER 1: INTELLIGENCE AS METAPHOR (AN INTRODUCTION)

16 **Why, for example:** Henry A. Kissinger, Eric Schmidt, and Daniel Huttenlocher, *The Age of AI: And Our Human Future* (New York: Little, Brown, 2021), 11–14.

CHAPTER 2: LETTER MAGIC

18 **In his epic history of:** Ibn Khaldun, *The Muqaddimah*, trans. Franz Rozenthal, vol. 1 of 3 (Princeton, NJ: Princeton University Press, 1958), 238–45.

19 **Ibn Khaldun wrote:** Khaldun, *Muqaddimah*, 1:242.

19 **The answers rather revealed a:** Khaldun, *Muqaddimah*, 1:245.

20 **"One should not think that":** Khaldun, *Muqaddimah*, 3:174.

20 **The question was then broken:** Khaldun, *Muqaddimah*, 3:213.

24 **partial table transcribed from *Ars Brevis*:** Ramon Llull, "Ars Brevis" in *Selected Works of Ramón Llull (1232–1316)* (Princeton, NJ: Princeton University Press, 1985), 581.

24 **Llull explained, that:** Llull, "Ars Brevis," 581.

25 **"Everything that exists,":** Llull, "Ars Brevis," 583.

27 **"The intellect banishes doubt":** Llull, "Ars Brevis," 596.

29 **The smart table, in the words:** Karl Marx, *Capital: A Critique of Political Economy*, trans. Samuel Moore and Edward Aveling (Chicago: Charles H. Kerr & Company, 1909), 82.

CHAPTER 3: SMART CABINETS

30 **Rumors of heresy circulated:** Blake Lee Spahr, "Quirin Kuhlmann: The Jena Years," *Modern Language Notes* 72, no. 8 (1957): 605–10.

31   **Under the spell of Llull:** German reference from Quirinus Kuhlmann, *Him-mlische Libes-Küsse*, ed. Birgit Biehl-Werner (Tübingen: De Gruyter, 1972). Translation by the author.

32   **Without the box, the young:** Synthesized from Quirini Kuhlmanni, *Epistolae Duae, Prior de Arte magna Sciendi sive Combinatoria (Cum Responsoria Athanasi Kircheri)*, printed by Lotho de Haes in Lugdunum Batavorum [Katwijk], 1674.

34   **Though his talents were recognized:** Robert L Beare, "Quirinus Kuhlmann: The Religious Apprenticeship," *PMLA* 68, no. 4 (1953): 828–62.

41   **The rules of grammar and:** Sidonie Clauss, "John Wilkins' *Essay Toward a Real Character*: Its Place in the Seventeenth-Century Episteme," *Journal of the History of Ideas* 43, no. 4 (1982): 531–53.

41   **"So that if men should generally":** John Wilkins, *An Essay Towards a Real Character, and a Philosophical Language* (London, 1668), Dedication; 20.

42   **"Now if these *Marks* can":** Wilkins, *Essay Towards*, 21.

43   **And though Wilkins could not:** Wilkins, *Essay Towards*, Epistle; 434.

44   **He called it *Plus Ultra*:** As quoted in Mogens Laerke (1686), "Leibniz, the Encyclopedia, and the Natural Order of Thinking," *Journal of the History of Ideas* 75, no. 2 (2014): 238.

45   **Without it, we are like:** Gottfried Wilhelm Leibniz, *Philosophical Papers and Letters*, ed. Leroy Loemker (Dordrecht, NL: Kluwer Academic Publishers, 1989), 222–24.

45   **For let the first terms:** Gottfried Wilhelm Leibniz, *Logical Papers: A Selection*, ed. George Henry Radcliffe Parkinson (Oxford: Oxford University Press, 1966), 11.

CHAPTER 4: FLORAL LEAF PATTERN

49   **The device, which he proudly:** Benjamin Woolley, *The Bride of Science* (New York: McGraw-Hill, 1999), 123–62.

50   **Also, did you know that:** Charles Babbage, "Street Nuisances" in *Passages in the Life of a Philosopher* (London: John Murray, 1864), 253–71.

51   **While other visitors yawned:** Sophia Elizabeth De Morgan, *Memoir of Augustus De Morgan* (London: Longmans, Green, and Co., 1882), 89.

52   **The switch to another ruleset:** Babbage, "Miracle" in *Passages in the Life*, 291–93.

53   **Frustrated with their slow progress:** Charles Babbage, "On a Method of

Expressing by Signs the Action of Machinery," *Philosophical Transactions of the Royal Society of London* 116, no. 1/3 (1826): 250–65.

57   **Whatever the arrangement of objects in:** L. F. Menabrea and Ada Lovelace (1842), "Sketch of the Analytical Engine Invented by Charles Babbage Esq." in *Scientific Memoirs, Selected from the Transactions of Foreign Academics of Science and Learned Societies, and from Foreign Journals*, ed. Richard Taylor (London: Richard and John Taylor, 1843), 666–731.

CHAPTER 5: TEMPLATE CULTURE

60   **Within the gears:** Menabrea and Lovelace, "Sketch of the Analytical Engine," 666–731.

64   **"It is here that the union":** Charles Babbage, *On the Economy of Machinery and Manufactures* (London: Charles Knight, 1832), 113.

66   **check my receipts:** Dennis Yi Tenen, "The Emergence of American Formalism," *Modern Philology* 117, no. 2 (November 2019): 257–83.

68   **Four subtypes were further enumerated:** Georges Polti, *The Thirty-Six Dramatic Situations*, trans. Lucile Ray (Ridgewood, NJ: Editor, 1917), 30.

69   **Further, rigorous experimentation was bound to change art:** Polti, *Thirty-Six Dramatic Situations*, 129–43.

77   **They were of limited horsepower:** Arthur Sullivant Hoffman, *The Writing of Fiction* (New York: Norton, 1934), 40.

CHAPTER 6: AIRPLANE STORIES

81   **A typical passage from the book:** Vladimir Propp, *Morphology of the Folktale*, trans. Laurence Scott (Austin, TX: American Folklore Society, University of Texas Press, 1968), 92.

83   **At a cocktail party:** Jean Piaget, *Structuralism*, trans. Chaninah Maschler (London: Routledge and K. Paul, 1971), 3–16.

84   **Roman Jakobson, who helped coin:** Roman Jakobson, *Selected Writings, II: Word and Language* (The Hague, NL: Mouton, 1971), 711.

87   **Victor Yngve, another prominent MIT linguist:** Victor H. Yngve, "Random Generation of English Sentences" (paper presented at the International Conference on Machine Translation of Languages and Applied Language Analysis, National Physical Laboratory, Teddington, UK, September 5–8, 1961), 66–80.

92   **"No more than about seven":** Yngve, "Random Generation," 66–80.

94   **The computer program complying Meehan's:** James Richard Meehan, "The Metanovel: Writing Stories by Computer" (PhD diss., Yale University, Department of Computer Science, September 1976).

CHAPTER 7: MARKOV'S PUSHKIN

102 **The word *oatmeal* will come:** C. K. Ogden and I. A. Richards, *The Meaning of Meaning: A Study of the Influence of Language Upon Thought and of the Science of Symbolism* (New York: Harcourt, Brace, 1936), 47.

103 **"As for my reasons":** L. I. Emelyakh, "Delo ob otluchenii ot tserkvi akademika A. A. Markova" [The case of the excommunication from the church of academician A. A. Markov], *Voprosy istorii religii i ateizma* 2 (1954): 397–411.

106 **"These semantic aspects of communication":** Claude Shannon, "A Mathematical Theory of Communication," *Bell System Technical Journal* 27, no. 3 (1948): 379.

110 **The program was capable of:** Charles M. Vossler and Neil M. Branston, "The Use of Context for Correcting Garbled English Text," in *Proceedings of the 1964 ACM 19th National Conference* (New York: Association for Computing Machinery, 1964), 42.401–42.4013.

CHAPTER 8: 9 BIG IDEAS FOR AN EFFECTIVE CONCLUSION

119 **In the conclusion to his book:** I. A. Richards. *Poetries and Sciences: A Reissue of* Science and Poetry *(1926, 1935) with Commentary* (New York: Norton, 1970), 76–78.

130 **In putting the algorithm in charge:** OpenAI, *GPT-4 Technical Report* (New York: arXiv, 2023).

INDEX

Reading Group Guide

Literary Theory for Robots
Dennis Yi Tenen

1. What does "writing well" mean to you? What standards or values are implicit in your judgment of what "good" writing is? Where and how did you learn those standards?

2. What tools or templates (paper or digital) do you use when reading and writing?

3. How does your word processor influence the way you write? What is being processed, and how is it processed?

4. How does technology influence creativity?

5. Who are the people and/or institutions responsible for designing a specific word-processing tool?

6. How is spell-checking implemented in your word pro-

cessor? What dictionaries are being used? Who wrote, edited, compiled, and/or published those dictionaries?

7. Do you think that the makers of these tools should be credited as "helping" your creative process?

8. What kinds of text generators do you encounter in the world? Think of specific institutional contexts, like medicine, law, or software engineering.

9. What ethical or political issues are raised in the seemingly autonomous operation of generative artificial intelligence (AI)?

10. How much of your own "original" work relies on existing templates you have found online or adopted from other sources? How do you know what is acceptable to use and what requires attribution and/or permission?

11. What role does smart technology play in the project of mass literacy? Where do various technologies, from "dictionary look-up" to auto spell-correction, lie on the spectrum of automation?

12. Who technically "writes" the output of a chatbot? Who do you think should be considered morally or legally responsible for something harmful that it does?

13. What are the potential consequences (social, cultural, or political) of smart technology?

14. At what point of "automation" does the technology seem "autonomous"?

15. What kinds of documents are being written by groups? Who do you think is responsible for quality in a group setting?

16. Where and how does textual technology operate in the world? Think of examples from various professional industries, including medical, legal, advertising, or even hospitality or aviation.

Complete Your Norton Shorts Collection:

Norton Shorts

BRILLIANCE WITH BREVITY

W. W. Norton & Company has been independent since 1923, when William Warder Norton and Mary (Polly) D. Herter Norton first published lectures delivered at the People's Institute, the adult education division of New York City's Cooper Union. In the 1950s, Polly Norton transferred control of the company to its employees.

One hundred years after its founding, W. W. Norton & Company inaugurates a new century of visionary independent publishing with Norton Shorts. Written by leading-edge scholars, these eye-opening books deliver bold thinking and fresh perspectives in under two hundred pages.

Available Winter 2024

Imagination: A Manifesto by Ruha Benjamin

Wild Girls: How the Outdoors Shaped the Women who Challenged a Nation by Tiya Miles

Against Technoableism: Rethinking Who Needs Improvement by Ashley Shew

Literary Theory for Robots: How Computers Learned to Write by Dennis Yi Tenen